A Practical
Guide to Content
Delivery Networks

OTHER AUERBACH PUBLICATIONS

Agent-Based Manufacturing and Control Systems: New Agile Manufacturing Solutions for Achieving Peak Performance
Massimo Paolucci and Roberto Sacile
ISBN: 1574443364

Curing the Patch Management Headache
Felicia M. Nicastro
ISBN: 0849328543

Cyber Crime Investigator's Field Guide, Second Edition
Bruce Middleton
ISBN: 0849327687

Disassembly Modeling for Assembly, Maintenance, Reuse and Recycling
A. J. D. Lambert and Surendra M. Gupta
ISBN: 1574443348

The Ethical Hack: A Framework for Business Value Penetration Testing
James S. Tiller
ISBN: 084931609X

Fundamentals of DSL Technology
Philip Golden, Herve Dedieu, and Krista Jacobsen
ISBN: 0849319137

The HIPAA Program Reference Handbook
Ross Leo
ISBN: 0849322111

Implementing the IT Balanced Scorecard: Aligning IT with Corporate Strategy
Jessica Keyes
ISBN: 0849326214

Information Security Fundamentals
Thomas R. Peltier, Justin Peltier, and John A. Blackley
ISBN: 0849319579

Information Security Management Handbook, Fifth Edition, Volume 2
Harold F. Tipton and Micki Krause
ISBN: 0849332109

Introduction to Management of Reverse Logistics and Closed Loop Supply Chain Processes
Donald F. Blumberg
ISBN: 1574443607

Maximizing ROI on Software Development
Vijay Sikka
ISBN: 0849323126

Mobile Computing Handbook
Imad Mahgoub and Mohammad Ilyas
ISBN: 0849319714

MPLS for Metropolitan Area Networks
Nam-Kee Tan
ISBN: 084932212X

Multimedia Security Handbook
Borko Furht and Darko Kirovski
ISBN: 0849327733

Network Design: Management and Technical Perspectives, Second Edition
Teresa C. Piliouras
ISBN: 0849316081

Network Security Technologies, Second Edition
Kwok T. Fung
ISBN: 0849330270

Outsourcing Software Development Offshore: Making It Work
Tandy Gold
ISBN: 0849319439

Quality Management Systems: A Handbook for Product Development Organizations
Vivek Nanda
ISBN: 1574443526

A Practical Guide to Security Assessments
Sudhanshu Kairab
ISBN: 0849317061

The Real-Time Enterprise
Dimitris N. Chorafas
ISBN: 0849327776

Software Testing and Continuous Quality Improvement, Second Edition
William E. Lewis
ISBN: 0849325242

Supply Chain Architecture: A Blueprint for Networking the Flow of Material, Information, and Cash
William T. Walker
ISBN: 1574443577

The Windows Serial Port Programming Handbook
Ying Bai
ISBN: 0849322138

AUERBACH PUBLICATIONS

www.auerbach-publications.com
To Order Call: 1-800-272-7737 • Fax: 1-800-374-3401
E-mail: orders@crcpress.com

A Practical Guide to Content Delivery Networks

Gilbert Held

Auerbach Publications
Taylor & Francis Group
Boca Raton   New York

Published in 2006 by
Auerbach Publications
Taylor & Francis Group
6000 Broken Sound Parkway NW, Suite 300
Boca Raton, FL 33487-2742

International Standard Book Number-10: 0-8493-3649-X (Hardcover)
International Standard Book Number-13: 978-0-8493-3649-2 (Hardcover)
Library of Congress Card Number 2005044212

This book contains information obtained from authentic and highly regarded sources. Reprinted material is quoted with permission, and sources are indicated. A wide variety of references are listed. Reasonable efforts have been made to publish reliable data and information, but the author and the publisher cannot assume responsibility for the validity of all materials or for the consequences of their use.

Library of Congress Cataloging-in-Publication Data

Held, Gilbert, 1943-
   A practical guide to content delivery networks / Gilbert Held.
       p. cm.
   Includes bibliographical references and index.
   ISBN 0-8493-3649-X (alk. paper)
   1. Computer networks. 2. Internetworking (Telecommunication) 3. Internet. I. Title.

TK5105.5H444 2005
004.6--dc22

                     2005044212

Taylor & Francis Group
is the Academic Division of T&F Informa plc.

Visit the Taylor & Francis Web site at
http://www.taylorandfrancis.com

and the Auerbach Publications Web site at
http://www.auerbach-publications.com

Dedication

Teaching graduate school courses in various aspects of data communications over the past decade has been a two way street with respect to the transfer of knowledge. Thus, it is with a considerable sense of gratitude for their inquisitive minds that I dedicate this book to the students at Georgia College and State University.

Table of Contents

Preface

The development of the World Wide Web resulted in a considerable effect on how we purchase items, review financial information from the comfort of our home, and the manner in which we perform work-related activities. Today, it is common for Web users to check their stock and bond portfolios, use a price comparison Web site prior to initiating a work-related purchase, check the latest results of their favorite sports team, examine weather predictions for a possible vacation getaway, and perhaps book airfare, hotel, and car rental all online. While we now consider such activities as a normal part of our day, what many may not realize is that our ability to perform such activities in a timely and efficient manner results from a hidden network within the Internet. That network and its operation and utilization are the focus of this book.

In this book, we will examine the role of content delivery networking to facilitate the distribution of various types of Web traffic, ranging from standard Web pages to streaming video and audio as well as other types of traffic. Because content delivery networking operations are normally performed by independent organizations, the appropriate use of facilities operated by different vendors requires knowledge of how content delivery networks operate. In this book, we will describe how such networks operate and the advantages and disadvantages associated with their utilization. In addition, we will examine Web architecture and the Transmission Control Protocol Internet Protocol (TCP/IP) because an appreciation of content delivery networking requires an understanding of both topics. Understanding Web architecture including the relationship between Web clients, servers, application servers, and back-end databases will provide us with the knowledge required to ensure the use of a content delivery network satisfies all of our organizational requirements. By understanding the TCP/IP protocol suite and the manner in which different applications are both transported and identified, we will obtain a base of knowledge that enables us to appreciate how content delivery networks operate. Because we can learn from the past, this author believes the evolution of technology represents an important aspect of any technology-related book. This book is no exception, thus, we will examine the development of a variety of technologies that evolved over the past decade as mechanisms to distribute various types of Web content. By understanding the advantages and the disadvantages associated with different types of content distribution technologies, we will obtain an appreciation for how we can use the networks within the Internet operated by various Internet Service Providers as a mechanism to deliver our organization's Web server–based information in a timely and efficient manner to both actual and potential users located around the globe.

As a professional author of technology-related topics I highly value reader feedback. You can either write me via my publisher whose mailing address is in this book or you can directly send me an email at gil_held@yahoo.com. Let me know if I spent too much or too little effort covering a particular topic, if I should have included another aspect of content delivery networking in this book, or any other comments you wish to share with me. Because I frequently travel I may not be able to respond to you overnight, but I will make every effort to respond to your comments within a reasonable period of time. Because many previous comments and suggestions concerning other books I wrote made their way into subsequent editions, it's quite possible that your comments will have a role in shaping the scope of coverage of a future edition of this book.

Gilbert Held
Macon, GA

Acknowledgments

The creation of the book you are reading represents a team effort even though there is only the name of this author on its binding and cover. From the acceptance of a proposal to the creation of a manuscript, from the proofing of the manuscript to the printing of galley pages, and from the correction of galley page typos and errors to the creation of the cover art and the printing of this book, many individuals contributed a considerable amount of time and effort. Thus, I would be remiss if I did not acknowledge the effort of several persons as well as the CRC Press publication team that resulted in the book you are reading. I am indebted to Rich O'Hanley, President of Auerbach Publishers, for agreeing to back another one of this author's research and writing projects. Similarly, this author is indebted to Claire Miller at Auerbach Publishers for her logistical support.

Due to a considerable amount of travel, this author realized many years ago that it was easier to write a book the old fashioned way using pen and paper rather than attempt to use a laptop or notebook. Pen and paper were easier when faced with circular, rectangular, square, and other oddball electrical receptacles that could vary from one side of a city to another. Using a few pens and a couple of writing pads was preferable to the uncertainty of the availability of applicable electrical plugs to power a portable computer. This author is indebted to his wife Beverly for her fine effort in converting his hand written chapters into a professional manuscript. In concluding this series of acknowledgments I would like to take the opportunity to thank all of the behind-the-scenes workers at Auerbach Publishers and CRC Press. From the creation of galley pages to printing, binding, and cover art, I truly appreciate all of your efforts.

chapter one

Introduction to Content Delivery Networking

The purpose of any introductory chapter is to acquaint readers with the topic or topics covered by a book. Commencing with a definition of a content delivery network (CDN) we will describe its evolution. We will examine several types of networking technologies that were developed to deliver specific types of content as well as the rationale for the development of the modern CDN.

1.1　The Modern Content Delivery Network

The modern CDN can be defined very simply as follows: A Content Delivery Network represents a group of geographically dispersed servers deployed to facilitate the distribution of information generated by Web publishers in a timely and efficient manner.

Although this definition is simplistic, it tells us that a CDN represents a group of servers that facilitate the distribution of information generated by Web publishers in a timely and efficient manner. To accomplish this task the servers must be located closer to the ultimate consumer or potential client operator than the server operated by the Web publisher.

Advantages

There are two key advantages associated with the modern CDN. Both advantages result from the fact that making multiple copies of the content of a Web publisher and distributing that content across the Internet removes the necessity for customer requests traversing a large number of routers to directly access the facilities of the Web publisher. This reduces traffic routed through the Internet as well as the delays associated with the routing of traffic. Thus, a CDN can be expected to reduce latency between client and server.

Disadvantages

Because the Internet is in effect a world wide network, a Web publisher can have hits that originate from almost anywhere on the globe. This means it would be impractical for almost all organizations to maintain duplicate servers strategically located around the world. Thus, most, if not all organizations need to rely on third party CDN operators that have globally distributed in-place networking equipment whose facilities are connected to the Internet. Most organizations need to use a third party. This results in disadvantages to include cost and support. There is also the need to ensure the CDN provider can update the Web publisher's server in a timely manner. Last, but far from the least, you need to verify the CDN provider has locations that as best as possible, correspond to the locations where potential groups or clusters of Web clients will access the organization's server. Now we have a general level of knowledge concerning what a CDN actually represents and know a few of the advantages and disadvantages associated with its use, let's probe deeper into this topic. We can consider the history factor whereby we learn from the past and examine the evolution of different content delivery methods.

1.2   Evolution

Although there were many technical problems associated with the development of the World Wide Web (www) series of servers created by academia, businesses, and government agencies, two problems were directly related to content delivery. Both problems resulted from the growth in the number of Web servers in which the total number of Web pages was increasing at an exponential rate.

The first problem resulted from the literal infinite availability of information, making it time consuming for individuals to locate information on the Internet. While search engines such as Yahoo! and Google facilitated the location of Web-based information, if a person required access to a series of data from different locations, he or she would need to spend an inordinate amount of time repeating the Web page location processes on a daily basis.

The second problem resulted from the global location of Web servers. Because a client in Chicago could require access to a server located in London, Web queries would need to flow thousands of miles through a series of routers to reach their destination. As each query reached the destination server in London, the computer there would process the request and return an applicable Web page. That Web page would flow in a reverse manner to the query, through a series of routers back to the client. Because routers need a small period of time to examine the header of each data packet as a mechanism to determine where to send the packet, network delays are primarily a result of the number of router hops between source and destination computers. As the number of hops between client and server increases so do the delays between the client query and server response.

In Chapter 2, we will obtain an appreciation for in depth details of client server computing. Until that chapter, we can logically note that the additional flow of traffic through the Internet and delays due to packets having to traverse a series of routers are impediments that adversely affect the flow of data. Of the two problems mentioned, the first resulted in the development of push technology. Because push technology contributed to the flow of data across the Internet and predated the establishment of CDN, we will discuss their basic operation as well as the different types of content delivery. However, prior to doing so lets refresh our knowledge by discussing the basics of client-server computing and note that it represents a pull technology.

Client-server computing

Although client-server computing represents a term we normally associate with the introduction of personal computers (PCs) during the 1980s, in actuality its origins date to the mainframe computers that were manufactured beginning in the 1950s and 1960s. We will briefly review data flow in the client to mainframe environment prior to examining modern client-server operations.

Client-to-mainframe data flow

By the mid-1960s, many mainframe computers had a hierarchical communications structure, using control units referred to as cluster controllers, to group the flow of data to and from a number of terminal devices cabled to each control unit. Control units were in turn either directly cabled to a channel on the mainframe computer when they were located in the same building as the mainframe, or communicated via a leased line to a communications controller when the mainframe was located in a different building or city. The communications controller, which was also referred to as a front-end processor, specialized in performing serial-to-parallel, parallel-to-serial data conversion as well as other communications-related tasks, offloading a majority of the communications functions previously performed by the mainframe. This enabled the architecture of the mainframe computer to be better designed for moving bytes and performing calculations. This in turn allowed the mainframe to process business and scientific applications more efficiently. In comparison, the communications controller was designed to process bits more efficiently, which represents a major portion of the effort required when taking parallel formed characters and transferring them bit-by-bit onto a serial line or conversely, receiving a serial data stream and converting the data stream bit-by-bit into a parallel formed character that the mainframe would operate on.

Figure 1.1 illustrates the hierarchical structure of a mainframe-based network. If you think of the terminal operators as clients and the mainframe as a server, since it provides access to programs in a manner similar to a modern day server, then the terminal to mainframe connection could be considered to represent an elementary form of client-server computing.

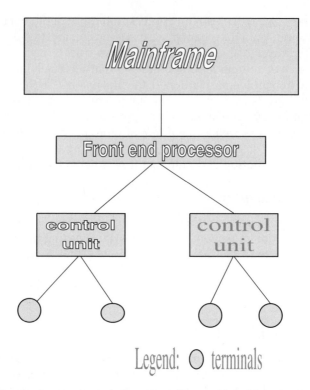

Figure 1.1 Terminal access to mainframes can be considered to represent an elementary form of client-server computing.

In the terminal to mainframe environment, the terminal functions as a client requesting a particular service. The mainframe, which functions as a server, represents the provider of the service. Because the first generation of terminal devices had a limited amount of intelligence, it wasn't until the second and third generation of terminals, which appeared during the later portion of the 1970s, that terminals were designed to perform what we would now refer to as limited functions, but were in effect hard-wired electronics. Those hard-wired electronics represented an early version of firmware and included a sequence of operations that could be considered to represent a program. Thus, the later generations of terminal devices were more representative of client-server computing than the first generation of terminals that simply displayed data and provided a limited data entry capability.

One of the key limitations of the mainframe environment was its original hierarchical structure. That is, all communications had to flow upward through the mainframe even if two terminal operators sat next to one another and desired to exchange information electronically. Although IBM attempted to change the hierarchical architecture of mainframes through its Advanced

Peer-to-Peer Networking (APPN) hardware and software introduced during the 1990s, by then local area networking was dominant to include modern client-server communications.

Modern client-server operations

Although we previously noted that the relationship between terminals and mainframes could be considered to represent an elementary form of client-server computing, it wasn't until the introduction of the PC during the early 1980s that this term gained acceptance. The programmability of PCs formed the basis for the development of modern client-server computing to include Web-based applications whose operations are improved by the use of a CDN.

In a modern client-server computing environment, the client represents a process (program) that transmits a message to a server process (program) over a communications network. The client process requests the server to perform a particular task or service. The client operates a program that normally manages the user interface portion of server programs, although it may also perform other functions. In a Web environment, the PC transmits Uniform Resource Locator (URL) addresses indicating the address of information the client wishes to receive. The server responds with Web pages that include codes to define how information on the pages should be displayed. The client program, which is the modern day browser, deciphers the imbedded codes to generate Web pages on the client's display.

Figure 1.2 illustrates the relationship of several aspects of modern day client-server computing in a Web environment. In the upper-right corner of the main window of the Microsoft Internet Explorer browser program, you will note the address http://www.yahoo.com. This address represents the URL transmitted to the Yahoo! server from the author's PC. In response to this query, the Yahoo! server transmitted the home page that corresponds to the URL address it received. In actuality, the client PC receives a sequence of Hyper Text Markup Language (HTML) and JAVA or ActiveX statements, a portion of which are illustrated in the window in the foreground that is located in the upper left portion of Figure 1.2. By selecting Source from the browser's View menu, the source statements that were interpreted by the PC operating the browser are displayed in a Notepad window. In this single display you can view the client request in the form of the URL transmitted to the server, and the server's response, which when interpreted by the browser operating on the client, generated the Web page display. In Chapter 2, we will describe HTML to include embedded programs coded in Java and ActiveX.

Because the Internet and its assorted facilities, such as Web servers, are recent additions to the history of client-server computing, it should be noted that servers perform other functions besides generating Web pages. Some servers execute database retrievals and update operations. Other servers provide access to printers, while a server-based process could operate on a

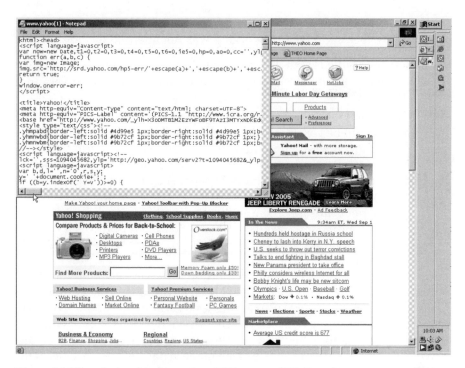

Figure 1.2 The relationship between a URL query, HTML code response, and client display of a requested Web page.

machine that provides network users access to shared files, resulting in a device referred to as a file server. Over the years, a number of different types of servers were developed to support a variety of applications. While a database server represents one common type of server, when the database is incorporated into a reservation system, the result is a specific type of reservation system server, such as a car rental, hotel, or airline reservation system server. Figure 1.3 illustrates a generic information delivery system where the network could be the Internet, a corporate intranet, or simply a local area network (LAN) without a connection to another network.

Figure 1.3 A generic information-delivery system.

Server network architecture

Similar to network architectures developed to expedite the flow of data, the use of servers resulted in an architecture being developed to facilitate processing. Initially, PC networks were based on the use of file servers that provided clients with the ability to access and share files. Because the file server would respond to a client request by downloading a complete file, the server's ability to support many simultaneous users was limited. In addition, network traffic occurring because of many file transfers could significantly affect communications. The limitations associated with file sharing resulted in the use of database servers. Employing a relational database management system (DBMS), client queries were directly answered, reducing network traffic considerably in comparison to the use of networks when a total file transfer activity occurred. The modern client-server architecture in many organizations results from the use of a database server instead of a file server.

One must consider two types of server architecture to expedite the flow of data. Those types of architecture are internal and external to the server. When we discuss internal server architecture, we reference the arrangement and specifications of the components within the server, to include processors, processor power, memory, channels to disk, disk capacity, and disk I/O data transfer rates. When we discuss external server architecture, we reference the subdivision of effort by having one server function as a preprocessor for another. Because this type of server relationship commonly occurs over a network, it's common to refer to this architecture as a server network architecture. In this section, we will focus on external or server network architecture. The two most common forms of server architectures are two-tier and three-tier.

Two-tier architecture

A two-tier architecture represents direct communications between a client and server, with no intervening server being necessary. Here the client represents one tier and the server represents the second tier. This architecture is commonly used by small to medium sized organizations where the server needs to support up to approximately 100 users.

Three-tier architecture

In a three-tier server architecture, a server or a series of servers function as agents between the client and the server where the data or the application they require resides. The agents can perform a number of functions that offload processing that otherwise would be required to be performed by the server. For example, agents could provide a translation service by placing client queries into a database retrieval language for execution on a database server. Other possible agent functions could range in scope from functioning

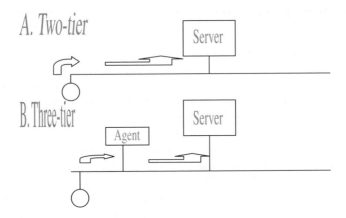

A. Two-tier

Server

B. Three-tier

Agent

Server

Figure 1.4 Two-tier versus three-tier architecture.

as a metering device that limits the number of simultaneous requests allowed to flow to a server to functioning as a preprocessor mapping agent, where requests are distributed to different servers based on certain criteria. Still other examples of functions that could be performed by middle tier devices include queuing of requests, request filtering, and a variety of application preprocessing that is only limited by one's imagination.

Figure 1.4 illustrates two-tier and three-tier architectures on a common network. In the two-tier architecture, note that client-server communications flows directly between devices. In a three-tier architecture data flow occurs twice on the network, first from the client to the agent and then from the agent to the server. In a LAN environment the additional network utilization needs to be compared with the offloading of processing functions on the server.

Now we have an appreciation for the basics of client-server architecture, let's return to our discussion on the evolution of content delivery and focus on push technology.

The road to push technology

Push technology is normally thought of as a new model of information distribution and data retrieval. In actuality, early versions of push systems occurred during the 1970s and early 1980s in the form of teletext systems and broadcast delivery videotext.

Teletext systems

Teletext systems began to become popular in the late 1970s, especially in Europe. Information pages are transmitted in the vertical blanking interval (VBI) of television (TV) signals, with decoders built into TV sets becoming capable of capturing, decoding, and displaying pages selected by the consumer. As many persons who travel to Europe probably remember, most

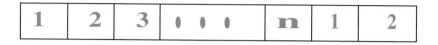

TV VBI data stream in
the form of 'pages' of
imformation

Figure 1.5 Teletext system operator.

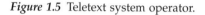

Figure 1.6 Optimal Teletext cycle where pages 1, 2, and 3 occurs 50, 30, and 20 percent of the time.

televisions in hotels include a teletext capability that enables a person holding a remote control to view the weather, TV schedule, and other information.

As illustrated in Figure 1.5, a teletext system transmits a sequence of information in the form of pages that are repeated at predefined intervals. The person with the remote enters a request, which results in the display of the desired information once it occurs in the broadcast stream.

Because teletext operators noted that certain pages of information were more desirable than other pages, they altered the sequence or the cycle of page transmissions from a pure sequence previously illustrated in Figure 1.5. For example, suppose there are only three pages of teletext information and the information on page one could be requested 50 percent of the time while the information on pages two and three might be desired 30 percent and 20 percent of the time, respectively. Then an optimum teletext cycle could appear as illustrated in Figure 1.6.

Videotext

While we can consider teletext as an elementary form of push and select technology, videotext added block formed pictures so the technology can actually be considered as the father of modern Web technology. Originally, videotext systems delivered information and transactional oriented services for banking, insurance, and shopping services. Later, videotext technology was used by newspaper publishers who made news and advertisements available through special terminals hooked up to television monitors. However, it wasn't until teletext systems were developed for operation with the growing base of PCs that teletext with block characters that provided what we would consider to represent elementary graphics today became widely available. By the mid-1990s prior to the availability of the Web, videotext systems were operated by America Online, CompuServe, Prodigy, and

Genie. In France, the French Government distributed Minitel terminals in place of telephone directories. Videotext is still in popular use in that country.

Pull technology

Until the introduction of push technology, information retrieval was based on what is referred to as pull technology. That is, a person would either have previously located an item of interest or used a search engine, such as Google or Yahoo! to locate an item of interest. Once an item of interest was located, the person, acting as a client operator, would use his or her browser to point to the URL on a server to retrieve the information of interest.

Role of caching

One popular method developed to facilitate client-server pull operations is caching. In a browser environment, which most readers are familiar with, caching occurs through the temporary storage of Internet files in a predefined folder on your hard drive. The stored files represent previously visited Web pages and files, such as graphics displayed on a particular page.

In a Microsoft Explorer browser environment, temporary Internet files are placed in a predefined folder located at: c:Documents and Settings\Owner\Local Settings\Temporary Internet Files\.

A browser user can go to Tools> General> Settings in Microsofts Internet Explorer to view cached files, adjust the amount of disk space to use for caching, change the location of the cache folder, and define how caching occurs.

Figure 1.7 illustrates the Settings Window selected from Tools> General> Settings on this author's computer. Note you have several options concerning the manner by which caching updates occur. You can have your browser check for newer versions of stored pages on every visit to the page, every time you start Internet Explorer, Automatically (browser default), or never. In addition, you can adjust the amount of space used for caching. While increasing the amount of disk space can increase how fast previously visited pages are displayed, it also results in additional processing since additional pages need to be searched. In addition, an increased disk cache obviously decreases the amount of space available for other files on the computer.

While Web caching is probably the most popular form of caching in use, it has certain strengths and limitations. Web caching is effective if documents do not change often. It becomes less effective as document changes increase in frequency. There are several deficiencies and limitations associated with pull technology and caching.

Pull limitations

To obtain information on a topic that is frequently updated, the user needs to periodically check the server. For example, during 2004, information about

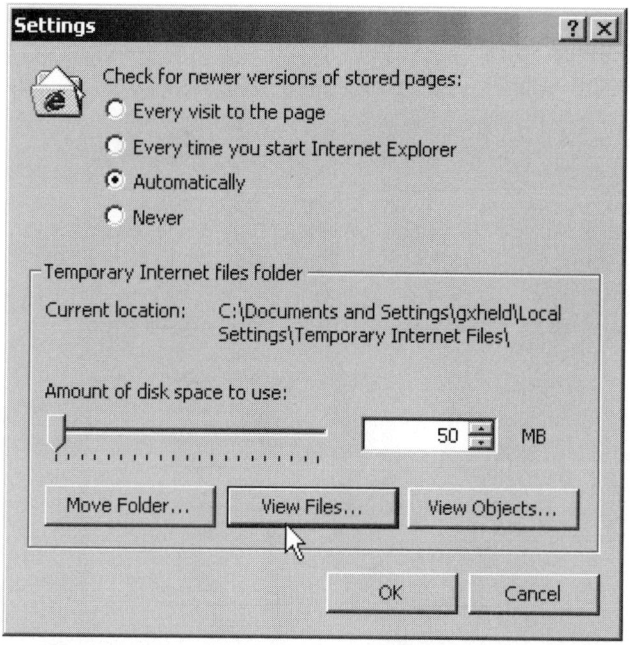

Figure 1.7 In Microsofts' Internet Explorer you can use Tools > General > Settings to control the use of caching.

hurricanes Charlie, Francis, and Ivan were of interest to persons living in the Southeastern United States. Under the pull model, users interested in hurricane information had to periodically query the United States National Hurricane Center Web site or a similar location, to obtain a forecast of the current and projected movement of the storm of interest.

To illustrate the preceding, let's assume you were interested in projecting the movement of Hurricane Ivan as it raced towards the United States during September 2004. Figure 1.8 illustrates the three-day projection for the movement of hurricane Ivan as of 11 A.M. on September 13, 2004. Obviously, if you lived in the projected path of this hurricane you would periodically have to pull a new projection to see if the path altered.

From a network viewpoint there can be hundreds to millions of persons accessing the same information because they are not sure when a new forecast will be available. This means the Web server may have to handle a volume of connections that can significantly vary. Depending on the location of clients with respect to the server as well as the amount of traffic flowing to clients latency can considerably increase. This may not only adversely affect persons accessing one or a few servers, it can also adversely affect other persons running different applications.

Because applications of interest have no efficient mechanism to avoid duplicated data, another limitation is the quantity of traffic in a pull environment. For example, suppose one user transmits a 256-byte packet representing

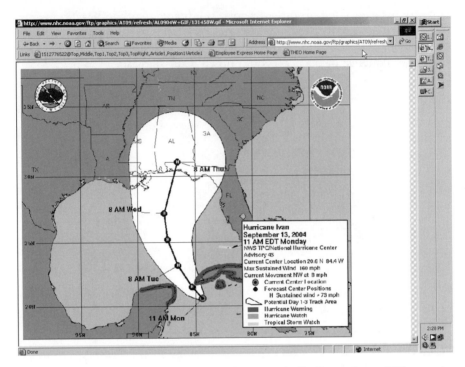

Figure 1.8 Using pull technology a client has to periodically retrieve a Web page to determine if an item of interest changed.

a URL request to a server that elicits a 10,000 1024-byte sequence of packet responses. If 1000 users requested the same information, the server must respond one thousand times with 10,000 1024-byte packets. During the 1980s and early 1990s several attempts were made to minimize the bandwidth required for the distribution of popularly requested information. The multi cast transmission is the most popular method still in use today.

Multicast

Multicast communications represents a method to conserve bandwidth in a Transmission Control Protocol/Internet Protocol (TCP/IP) environment. Multicast communications occurs between a single data source and multiple receivers. Although multicast represents a 1990s technology, it provides several important advantages for group communications in comparison to traditional unicast transmission.

Advantages

A primary advantage in using multicast is it enables the conservation of network bandwidth. For example, if ten clients residing on a network subscribe to a multicast video, only one sequence of packets flows onto the network instead of ten. Another important advantage of multicast transmission is it scales very well to large user groups. That is, if the number of clients on a network subscribe to the multicast video increases to 100 or even 1000, the same sequence of packets would flow on the network.

Addresses

In a TCP/IP environment the range of Internet Protocol Version 4 (IPv4) addresses from 224.0.0.0 thru 239.255.255.255 are reserved for multicasting. These addresses are also referred to as Class D addresses. Every IP datagram whose destination address commences with the first four bits set to 1110 represents an IP multicast datagram. The remaining 28 bits in the address identify the multicast group that the datagram is sent to.

Within the Class D address block, the range of addresses between 224.0.0.0 and 224.0.0.255 are reserved for use by routing protocols, topology discovery protocols, and maintenance protocols. Within this address range are several well known multicast groups. Two examples of these groups include:

```
* 224.0.0.1 is the all-hosts group.
```

If you ping this Class D address, all multicast capable hosts on the network will respond.

```
* 224.0.0.2 is the all-routers group.
```

All multicast routers must join that group on all its multicast capable interfaces.

Limitations

One of the major disadvantages of multicast is users must register to receive a multicast data stream, which makes this technology more suitable for pre-defined events such as video telecasts rather than pulling different information off several Web servers. Another limitation of multicast is not all routers or hosts conform to the multicast specification. In fact, there are three levels of conformance with respect to the multicast specification, with level 0 indicating no support, level 1 indicating support for sending but not receiving multicast datagrams, while level 2 indicates full support for IP multicast.

Push technology

Push technology was developed because of the need to bring information of interest to consumers instead of having them retrieve the data. Under push technology, a consumer signs up with a push provider to receive certain data transmitted over the Internet to the user's desktop. Information of interest is displayed either as a stream on the 25th line of the screen, or perhaps in a separate window on the display. The actual data sent to a user's desktop can come from a variety of Web sites depending on the user's interest. When the user fills out a profile of data of interest with the push provider, that profile functions as a filter. The vendor's server searches a variety of Web sites, collecting information of interest to all subscribers; however, it uses each subscriber profile as a mechanism to determine the data to push to individual subscribers. Because selected information of interest is broadcast similar to a television or radio broadcast the terms streaming, channeling, and broadcasting are sometimes used synonymously. According to some persons, the term push also derives from the term push polling that was used during the 1996 U.S. presidential election, when unscrupulous polling personnel pretended to conduct a telephone opinion poll but, in actuality, used questions that pushed their candidates' strengths.

Evolution

In a modern push technology environment, PointCast represents the vanguard of a series of companies that developed push technology. The company was founded in 1992 to deliver news and other information over the Internet. During 1996, the company distributed approximately 1.6 million copies of its proprietary Webcasting software, which for its time, represented a tremendous achievement. During 1997, the hype associated with push technology resulted in *Business Week* noting PointCast in its cover story on Webcasting to include a quotation from its then 34-year old CEO, Chris Hasset, that "We are defining a new medium."

PointCast's run at fame was momentary. PointCast provided customers with the ability to receive advertisements along with specific requested information. But push technology was shoved off corporate networks by organizations that found employees were using the PointCast system and were clogging their Internet connection as well as the local area networks connected to the mother of all networks. As more and more companies shunned the technology, the fortunes of PointCast considerably diminished. In addition, competition from Yahoo! and other Internet-related commercial firms resulted in using push technology via non-proprietary Internet channels. Eventually both Microsoft and Netscape incorporated Webcasting technology into their browsers, further diminishing the need for PointCast's proprietary software. Within a few years, another company acquired PointCast and push technology from the pioneer ceased.

The adoption of push technology was valuable for organizations needing a mechanism to deliver information *en masse*, such as automatic updating of business price lists, manuals, inventories, and policies. However, the mass market never actively used the desktop channel bar available under Internet Explorer Version 4.0. The more modern versions of Windows, such as Windows XP, do not support the desktop channel bar unless you upgrade from Internet Explorer 4.0 or Windows 98. Most persons use Microsoft's Windows Media Player, RealNetworks RealOne, or a similar program to view predefined audio and video. While you can use either program to retrieve news, play, burn, and rip audio tracks, and find a variety of content on the Internet, these modern Webcasting tools are more pullers than pushers. The modern Webcasting tools back one key feature built into many push products and was included in Internet Explorer Version 4.0. That feature was a crawl capability, which we will briefly discuss prior to moving on to modern content delivery methods.

Crawling

Under Internet crawling, a user specifies whether they want the general type of information or a specific type of information. The client program performs a site-crawl of applicable Web sites where it examines data for relevance, checks for updated content, then notifies the user of changes. Users of Internet Explorer 4.0 could specify Web crawls three levels deep from a subscribed Web page and retrieve pages from non-subscription sites.

To perform a crawl requires a specified starting Web page. Given that page, a browser crawls each link on that page and, if configured to do so, the browser can crawl multiple levels deep or restrict its operation to each specified page in a list. During the crawl operation, the browser will check to see if a page changed by comparing file size and creation date of the stored and access files. Depending on the configuration of the crawl, once a page is checked, either a notification of changes, or actual page content is downloaded to the user.

Advantages

Similar to most technologies, there are several advantages associated with push technology. From the viewpoint of the client, push technology permits predefined requests that enables items of interest to be received when such items have Web page changes. From the viewpoint of the server, push technology enables common multiple requests to result in a single response to the push provider, which in turn becomes responsible for the transmission of Web pages to individual client subscribers. Due to this, push technology is potentially extremely scalable and can reduce the load on both distant servers and networks, which in turn can be expected to enhance response time.

Disadvantages

Previously we noted that the main reason for push technology failing to live up to the hype resulted from its use clogging corporate networks. Besides clogging corporate networks, push technology had two additional limitations, which were its need for clients to subscribe to a service and control. Concerning subscription fees, most Internet users view the Internet as a free service and always want cheap or no-cost information. Concerning control, most users prefer on-demand information retrieval (pull) where they control information to be downloaded. Because push technology was developed in an era where high speed LANs operated at 10 Mbps, the traffic resulting from push technology had a significant effect on corporate networks. When combined with the subscription nature of most push operators and user preference for control of Web page retrieval, it's a wonder push technology held the interest of Internet users during the 1990s.

Now we have a general level of appreciation for the evolution of a variety of information retrieval techniques that were developed over the past half century, let's examine the role of modern content delivery networking and how it facilitates many types of client-server communications to include push, pull, and Web crawling.

1.3 Content Delivery Networking

We were introduced to content delivery networking through an abbreviated definition of what the term means. We were introduced to the evolution of client-server technology including caching, pull, push, and crawling operations. In this section, we will use this knowledge to appreciate the manner by which content delivery networking facilities delivers the various types of client-server operations. Because the benefits obtained from using a CDN requires some knowledge of the limitations of client-server operations and the general structure of the Internet, let's begin our examination of CDN by discussing client-server operations on the Internet.

Client-server operations on the Internet

The Internet represents a collection of networks tied to one another using routers that support the TCP/IP protocol suite. To illustrate some of the problems associated with content delivery as data flows across the Internet, let's first assume a best case scenario where both client and server are on the same network. Then, we can expand the distance between client and server in terms of router hops and networks traversed, introducing the point of presence (POP), and peering point used to interconnect separate networks to one another on the Internet.

Client-server operating on the same network

When we speak of the Internet, the term network represents a collection of subnets with connectivity provided by an Internet Service Provider (ISP). When we mention a client and a server reside on the same network on the Internet, the two computers can reside on the same segment, or on different subnets that require one or more router hops to be traversed for one computer to communicate with the other.

The delay associated with the delivery of server content primarily represents a function of network traffic, available network bandwidth, and router hops from client to server. Because an ISP controls the network, content delivery is more manageable than when data delivery has to flow between interconnected networks. As a mechanism to minimize the effect of additional traffic, the ISP can upgrade bandwidth and routers as they sign up additional clients.

Client-server operations on different networks

When the client and server are located on different Internet networks, traffic must flow through an access point where networks operated by different ISPs are interconnected. On the Internet, a POP is sometimes used as a term to reference an access point where one network connects to another. In actuality, the term POP has its roots in telephony. POP originally represented the physical location where the local telephone operator connected the network to one or more long distance operators. While the term POP is still used to reference the location where two ISPs interconnect their networks, a more popular term used to reference this location is the Internet peering point.

Peering point

An Internet peering point represents the physical location where two or more networks are interconnected. Such locations are based on contractual agreements between ISPs and trace their origins to the original expansion of Advanced Research Project Agency Network (ARPANET). As the Internet evolved, ARPANET was considered to represent a backbone network, with other networks linked to one another via one or more connections to the backbone. As the Internet expanded and evolved, there is no longer a single backbone network in the traditional meaning. Instead, various commercial ISPs as well as private network operators entered into agreements whereby two or more networks were interconnected at a peering point under a peering agreement. Today there are two main types of peering — private and public. A private peering point results in an agreement between two ISPs to permit traffic to flow between two networks. In comparison, a public peering point, also referred to as an Internet Exchange Point, represents a location independent of any single provider where networks can be interconnected. ISPs with large traffic volumes, such as MCI (formerly known as WorldCom) are often referred to as Tier 1 carriers and usually establish peering agreements with other Tier 1 carriers without charging one another for the interconnection. Smaller providers with lighter traffic loads tend to use Internet Exchange Points where they pay a fee for interconnection services.

One example of a peering point is MAE-EAST. The term MAE stands for Metropolitan Area Ethernet and represented an interchange constructed by Metropolitan Fiber Systems (now owned by MCI) for PSI, UUNET, and SprintLink during 1993. MAE-EAST was established at approximately the same time the National Science Foundation was exiting the Internet backbone business. This enabled the peering location to become so successful a similar facility was opened in Silicon Valley, referred to as MAE-WEST. By 2005, MAE-EAST had expanded to four sites in the Washington, DC, metropolitan area and one location in New York City, with 38 members ranging in size from AT&T WorldNet and BT, to Epoch Networks, Equant, Hurricane Electric, Infornet, Swiss Com AG, UUNET, Verio, and Xspedius.

Today there are three major MAEs in the United States to include MAE-EAST, MAE-WEST, and MAE-Central, with the later located in Dallas, Texas. In addition, two central MAEs for frame encapsulation (FE) service are located in Chicago and New York.

Figure 1.9 illustrates the Internet Traffic Report for MAE-EAST. Note that Figure 1.9 illustrates two graphs, each indicating activity for the past 24 hours. The top graph indicates a traffic index, which represents a score from 0 to 100, where 0 is slow and 100 is fast. The traffic index is computed by comparing the current response of a ping echo to all previous responses from the same router over the past seven days, with a score of 0 to 100 assigned to the current response depending if this response is better or worse than all previous responses from the route.

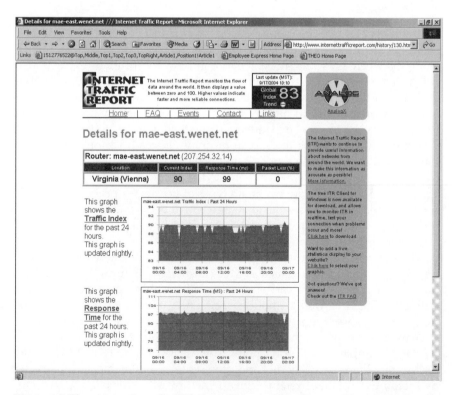

Figure 1.9 The top portion of a Web page that provides three metrics concerning the operation of MAE-EAST.

The second graph indicates response time in milliseconds (ms). The response time represents a round trip computed by sending traffic from one location to another and back. If you scroll down Figure 1.9 you view a third graph labeled packet loss. That graph indicates the percent of packets dropped by the router or otherwise lost. Typically, routers discard packets when they become overloaded. This represents a measurement of network reliability.

The three metrics displayed by the Internet Traffic Report partially shown in Figure 1.9 represent bottlenecks when information flows between ISP networks. The peering point can be viewed as a funnel through which all traffic from one Internet network destined to a different network must flow. Because the flow of data between ISPs is usually not symmetrical, some ISPs may have more data to transfer through a peering point than the connection can handle. When this situation occurs, the connection becomes a bottleneck. Although the peering point could be upgraded, only a few customers operating servers experience a problem, and the ISP where the server resides may very well be reluctant to upgrade the peering point due to the cost involved. Similarly, the installation of additional peering points can represent a significant cost. Even if both existing peering points are upgraded and additional peering points established to provide extra inter-network connectivity, doing so takes time and may not appreciably decrease delays currently experienced by clients on one network attempting to access servers on another. The latter situation results in a fixed number of router hops needing to be traversed when a client on one network needs to access information from a server residing on another network.

Because of the previously mentioned problems, another solution evolved during the 1990s. This solution was based on moving server content from a central location to multiple distributed locations across the Internet. In doing so Web pages were moved closer to the ultimate client requester, reducing both the number of router hops a request would have to traverse as well as the round-trip propagation time. In addition, since Web pages are now closer to the end-user, traffic would not have to flow through peering point bottle-necks, enhancing traffic flow and minimizing traffic delays. The effort involved in distributing Web pages so they are available at different locations is known as content delivery networking. The role of the modern CDN is to facilitate the retrieval of Web-based information by distributing the information closer to the ultimate requestor. In doing so CDN facilitates a variety of client-server communications to include pull, push, and Web crawling.

chapter two

Client-Server Models

In the first chapter of this book, we considered client-server architecture to represent a single model that varied by the way data flowed between each computer. If data flowed directly from the client to the server, the architecture could be represented as a two-tier architecture, with the client representing the first tier and the server representing the second tier. If data flowed from the client to the server and depending on the request, flowed to another server, we could refer to the architecture as being a three-tier architecture. As a review, in a two-tier architecture the user interface is typically located on the user's desktop while the database is located on a server that provides services to many clients. In a three-tier (also referred to as multi-tier) architecture, a middle layer is added between the client and the database to be accessed. The middle layer can queue requests, execute applications, provide scheduling, and prioritize work in progress. Because a number of different types of software products can reside at each tier, this can result in a series of different client-server models and results in the title of this chapter.

In this chapter, we will turn to a core set of software products that operate on the client and the server. The interaction of such products results in different client-server models, with each model having varying characteristics that are affected by latency as data moves across the Internet, as well as traffic on each Internet Service Provider (ISP) network, and traffic routed through any points of presence (POP) as data flows from a client located on one ISP network to a server located on a different ISP network and a response that flows back to the client.

To begin our examination of client-server models we will follow a familiar tune and begin at the beginning. We will examine the three tiers that can be employed in different client-server architectures. We will note different types of popular software that can operate on each tier. We will use the information as a foundation to probe deeper into the characteristics of the different software on each tier, including their relationship with other software and the effect on software operations as the distance between client and server increases from residence on a common network to computers located on different networks.

2.1   Overview

Figure 2.1 illustrates the three tiers associated with modern client-server architecture. In this block diagram, potential software programs are indicated with respect to the tier where they would normally reside. In addition, common operating systems used at each tier are indicated to provide readers with additional information concerning platforms that are commonly used in the modern client-server environment.

2.2   Client Operations

In the wonderful world of the Internet, the browser represents the client. The purpose of the browser is to enable users to request documents from a server and to display the response. While Netscape Corporation developed the first commercially available browser, Microsoft Corporation's Internet Explorer now dominates the browser market, with approximately 90 percent of the market share. Other browsers such as Netscape, Firefox, Opera, and other products cumulatively hold the remainder of the market.

Browsers differ in the version of the Hyper Text Markup Language (HTML) they support and in their support of code modules, plug-ins, the amount of customization users can perform, and caching capability. Because of the market share of Microsoft's Internet Explorer, we will focus on this browser when discussing client operations.

Uniform Resource Locators (URLs)

Uniform Resource Locators (URLs) are short strings that identify the location of various resources on the Web. Those resources can include documents, images, downloadable files, electronic mailboxes, and services. The general format of a URL is as follows:

Protocol://location

Two common examples of protocol are Hyper Text Transfer Protocol (HTTP) and File Transfer Protocol (FTP). The location can include a significant degree of variety depending on where information resides. For example, the home page of the Web server whose domain is popcorn.com would be accessed as follows:

http://www.popcorn.com

Not shown in the above URL example, is the port number used for HTTP. By default, the port number is 80. If for some reason the port on the server being accessed uses a different port number, then the URL would become:

http://www.popcorn.com:number

where number represents the port number.

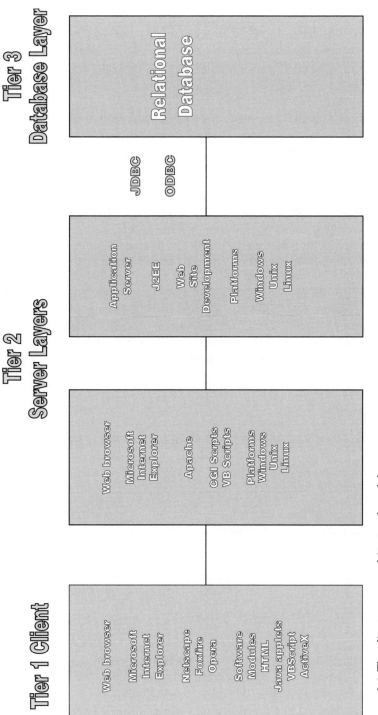

Figure 2.1 The client-server architectural model.

This URL would take you to the home page of the Web server whose address is www.popcorn.com and is configured to receive HTML queries on port 80. We should note we could also specify a path to a particular location on a computer, pass parameters to scripts via a query string, and even refer to a specific section within an identified resource using a fragment. Thus, a more detailed URL format would be as follows:

Protocol://domain name[IP address]:[port][path][Query][fragment]

As an example of an expanded URL, let's assume we want to log onto the home page of the Smith Barney financial Web site. You could point your browser to the Smith Barney home page and select a menu entry that will prompt you to enter your user information, or go directly to the access log-on page. To go directly to the access log on the home page, you would enter the URL as https://www.smithbarney.com/cgi-bin/login/login.cgi

From this URL note that the protocol was changed to HTTPS, where the 'S' stands for secure. Also, the Common Gateway Interface (CGI) represents a set of rules that describes how a Web server communicates with other software that can be on the client and the server. In our example, the Smith Barney access page will display a small form requiring you to enter your UserID and password. The CGI program will process the entered data. Later in this chapter, we will take a more detailed look at CGI.

To specify a path to a particular document, file, or service on a server, results in using one or more additional forward slash (/) following the server address and optional port number. Within the URL, a variety of special characters can be used to effect different operations. For example, the question mark (?) can be used to delimit the boundary between the URL of a queryable object, and a set of words used to express a query on that object.

Hyper Text Markup Language (HTML)

The document typically requested using a URL is usually encoded in HTML. HTML represents a markup language that defines a structure for a document without specifying the details of the layout.

HTML documents are plain text American Standard Code for Information Interchange (ASCII) files created using a text editor or word processor that enables text to be saved in ASCII format. The fundamental component of the structure of a text document is referred to as an element. Examples of elements include heads, tables, paragraphs, and lists. Tags let you mark the elements of a file for a browser to display. Tags consist of a left angle bracket (<), a tag name, and a right angle bracket (>); for example, <tagname>. Tags are usually paired with the termination tag prefixed with a backward slash (/), so <h1> and </h1> would be used to surround a header level 1 name. Table 2.1 has an elementary example of a hypertext-coded document.

Table 2.1 A Brief HTML Document

```
<html>
<head>
<title>
Title of  document
</title>
</head>

<body>
<h1> Header  level  1  </h1>
Some  text  goes  here
<h2> Header  level  2  </h2>
Some  text  goes  here
</body>
```

Hyper Text links

The Web represents a set of network accessible information resources constructed on three basic ideas. These ideas include URLs, which provide a
global location and naming scheme for accessing resources, protocols such
as HTTP for accessing resources, and hypertext in the form of HTML that
enables one document to contain links to other resources. Under HTML, a
unidirectional pointer referred to as an anchor is used in a tag along with a
URL to function as a hypertext link. The letter 'a' is used in an anchor tag
as an abbreviation for anchor, with the format for a hypertext link as follows:

```
<a href = destination>label </a>
```

where the <a> and </a> tags represent an anchor, destination represents a
URL, label represents the information displayed by the browser, which is
highlighted, and when selected, results in the browser retrieving and displaying information from the destination. The destination is hidden until
you move the cursor over the label. The URL hidden in the anchor will be
displayed on the bottom line of the browser. One example of using an anchor
is as follows:

```
See:<a href = www.popcorn.com/tasty> for tasty popcorn
</a> information
```

Hyper Text Transport Protocol (HTTP)

HTTP represents a stateless, connectionless, reliable protocol. HTTP is used
to transfer Web pages from a Web server to a client's Web browser using

Transmission Control Protocol (TCP), usually on port 80. The term stateless means each HTTP transmission occurs without needing information about what previously occurred. The protocol is connectionless, which means an HTTP message can occur without establishing a connection with the recipient. HTTP is a reliable protocol because it uses the TCP transport protocol, which provides a reliable error detection and correction facility. In Chapter 3, we will probe deeper into Transmission Control Protocol Internet Protocol (TCP/IP). For now, we will note that HTTP is transported by TCP within an IP datagram.

Versions

The current version of HTTP is 1.1, with previous versions being noted as 0.9 and 1.0. The first line of any HTTP message should include the version number, such as HTTP 1.1.

Operation

HTTP messages are stateless, connectionless, and reliable. HTTP messages fall into three broad categories: Request, Response, and Close.

Request message

Every HTTP interaction between client and server commences with a client request. The client operator enters a URL in the browser, by either clicking on a hyperlink, typing the URL into the browser address field, or by selecting a bookmark. Because of one of the preceding actions, the browser retrieves the selected resource. To accomplish this task the browser creates an HTTP request as shown below:

```
Request Line   GET/index.html HTTP/1.1
Header Fields  Host: www.popcorn.com
               User-Agent: Mozilla/4.0
```

In this HTTP request, the User-Agent represents the software that retrieves and displays the Web content. Mozilla identifies Netscape's browser, while Microsoft's Internet Explorer would have the string MSIE and version number placed in the User-Agent field.

In addition to the GET request, a browser can issue several other types of requests. Let's turn to the structure of HTTP request methods to include the format of HTTP Request methods.

A Request message is transmitted from a client to a server. The first line of the message includes the request method to be applied to the resource, the identifier of the resource, and the protocol version in use. To provide backward compatibility with prior versions of HTTP, there are two valid formats for an HTTP request, both of which are in Table 2.2.

Table 2.2 HTTP Request Formats

```
Request=Simple Request/Full-Request
Simple-Request="GET" SP Request-URL CRLF
Full-Request= Request-Line
              *(General-Header
              | Request-Header
              | Entity-Header)
              CRLF
              [Entity-Body]
```

In examining the HTTP Request message formats shown in Table 2.2, several items warrant discussion. First, if an HTTP 1.0 server receives a Simple-Request, it must respond with an HTTP 0.9 Simple-Response. An HTTP 1.0 client capable of receiving a Full-Response should never generate a Simple-Request. Secondly, the Request-line begins with a request method token, followed by the Request-URL and the protocol version, ending with a carriage return, line feed (CRLF). Thus, for a Full-Request, Request-line=Method SP Request-URL SP HTTP-Version CRLF.

The Method token identifies the method to be performed on the resource identified by the Request-URL, where under HTTP 1.0:

```
Method="GET"/"HEAD"/"POST"/extension method
```

and,

```
extension-method=token
```

The GET token is used to retrieve information identified by the Request-URL. The HEAD token functions similar to the GET, however, the server does not return any Entity-Body (information) in the response only HTTP headers are returned. The POST token provides a way to annotate existing resources, post a message, or provide a block of data to a server, such as submitting a form.

When the client transmits a Request, it usually sends several header fields. Those fields include a field name, a colon, one or more space (SP) characters, and a value. After the Request-line and General-Header, one or more optional HTTP Request-Headers can follow that are used to pass additional information about the client, its request, or to add certain conditions to the request. The format of a header field line is shown below:

Field Name	Value
Content-type	text/html

Table 2.3 lists seven common HTTP request headers to include a brief description of each.

Table 2.3 Common HTTP Request Headers

Header	Description
HOST	Specifies the target hostname
Content-length	Specifies the length (in bytes) of the request content
Content-type	Specifies the media type of the request
Authentication	Specifies the username and password of the user
Refer	Specifies the URL that referred the user to the current resource
User-agent	Specifies the name, version, and platform of the client
Cookie	Returns a name/value pair set by the server on a previous response

One of the more interesting aspects of an HTTP Request is the Referrer header field. If you browse a Web page and click on an anchor, the Referrer header field informs the destination server of the URL (think page being viewed) from where you invoked the anchor. This information can be used to determine indirect traffic flow and the effect of advertising.

Response message

After receiving and interpreting a Request message, a Web server replies with an HTTP Response message. The format of an HTTP Response message is shown in Table 2.4.

Similar to a Request message, a Simple-Response should only be returned in response to an HTTP 0.9 Simple-Request. The Status-line is the first line of a Full-Response message and includes the protocol version, followed by a numeric status code and its associated textual phrase, with each element separated by SP (space) characters. The format of the Status-line is:

```
Status-line=HTTP-Version SP Status-Code SP
Reason-Phrase CRLF
```

Table 2.4 HTTP Response Formats

```
Response = Simple-Response/Full-Response
Simple-Response = [Entity-Body]
Full-Response = Status-line
               *(General -Header
               | Response-Header
               | Entity-Header
               CRLF
               [Entity-Body]
```

Table 2.5 Defined Response
Status-Codes and Reason Phrase

Status-Code	Reason-Phrase
200	OK
201	Created
202	Accepted
204	No content
301	Moved permanently
302	Moved temporarily
304	Not modified
400	Bad request
401	Unauthorized
403	Forbidden
404	Not found
500	Internal server error
501	Not implemented
502	Bad gateway
503	Service unavailable

Table 2.5 lists defined Status codes and their reason phrase.

Using status codes can reflect multiple situations. For example, if a client went to a restricted Web server location, the server would reject the request with a 401 message. However, if the server wishes the client to authenticate its request it would reject the request with a 401 message, and show in the www-Authenticate field, information about the authentication requirements so the client can determine if it has authorization to authenticate. If it does, it would include its UserID and password in the request.

Hyper Text Transfer Protocol (HTTP) 1.1

The most popular version of HTTP is 1.1, which includes several improvements over prior versions. Some of the improvements include chunked data transfers, support for persistent connections that reduce TCP overhead, byte ranges that enable portions of a document to be requested, hostname identification that allows virtual hosts, content negotiation that permits multiple languages, and proxy support. While HTTP 1.1 is more efficient than prior versions of the protocol, it is also more complex. For example, the number of HTTP Request Methods is now increased to eight. Table 2.6 lists the

Table 2.6 HTTP 1.1 Request Methods

Method	Description
GET	Asks the server for a given resource and no content
HEAD	Similar to GET, but only returns HTTP headers and no content
POST	Asks the server to modify information stored on the server
PUT	Asks the server to create or replace a resource on the server
DELETE	Asks the server to delete a resource on the server
CONNECT	Used to allow SSL connections to tunnel through HTTP connections
OPTIONS	Asks the server to list the request methods available for a given resource
TRACE	Asks the server to echo back the request headers as it receives them

expanded set of Request Methods supported by HTTP 1.1 including a brief description of each method.

State maintenance

The HTTP protocol is stateless meaning an HTTP session only lasts from a browser request to the server response, after which any subsequent request is independent of the prior request. The stateless nature of HTTP represents a problem if you use the browser to perform activities where information needs to be maintained, such as selecting items from an e-commerce Web site to fill a shopping cart.

There are two solutions: one is using cookies, which are the most common method used to overcome the stateless nature of HTTP sessions. The second is using a hidden field in an HTML form whose value is set by the server.

Cookies

A cookie is a short file that functions as an identifier. The cookie is created by a Web server accessed by the client as a way for the server to store information about the client. The information can be its preferences when visiting the site or the items selected by the user for potential purchase.

Cookies are stored on the client for a predefined time. Each time a client transmits a request to a server any cookie previously issued by the server is included in the client request and can be used by the server to restore the state of the client. From a security perspective, once a cookie is saved on a client it can only be transmitted to the Web site that created the cookie.

If you're using Microsoft's Internet Explorer, you can view your browser's cookies by selecting Tools> Internet Options and select the Settings button in the Temporary Internet Files area and the View Files button in the resulting settings window. Figure 2.2 illustrates some of the cookies stored

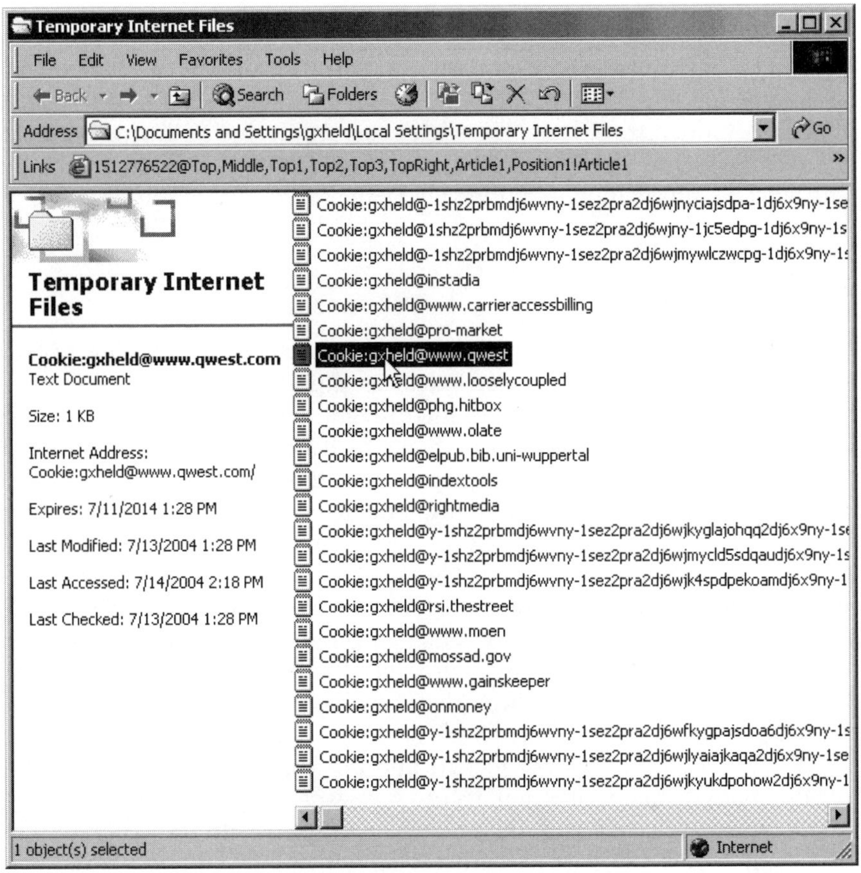

Figure 2.2 You can view cookies on your computer via Microsoft's Internet Explorer tools menu.

on the author's computer, with a cookie from Quest highlighted. Note that the left portion provides a description of the cookie to include its Internet address, when it expires, and when it was last checked, accessed, and modified.

Types of cookies

There are two types of cookies: persistent and temporary. A persistent cookie is one stored as a file on your computer, which remains stored when you close your browser. A persistent cookie can only be read by the Web site that created it when you access that site again. In comparison, a temporary cookie is only stored for your current browsing activity and is deleted when you close your browser. While browsers enable users to accept, prompt users for acceptance, or block cookies, when cookies are blocked, certain activities, such as filling a shopping cart may be difficult or impossible to achieve.

Hidden fields

A second method that can be used to overcome the stateless nature of HTTP sessions is hidden fields in an HTML form. The hidden field, as its name implies, is hidden from view. The server can set a value in the hidden field of a form, which when submitted by the client is returned to the server. By placing state information in hidden fields, the server can restore the state of the client.

Browser programs

The modern browser could represent a sophisticated mini-operating system that allows other programs, referred to as plug-ins, to operate under its control. The browser can also run interpreters that execute JavaScript and Visual Basic Scripting (VBScript), both of which are commonly used for field validation of forms. Another interpreter run within a browser is Java, which represents a high-level programming language that provides much more capability than Java Script or VBScript. One additional type of program that is unique in a Windows environment is an ActiveX control. An ActiveX control represents a dynamic link library (DLL). In the wonderful world of computers, a DLL represents a collection of small programs, any of which can be invoked when needed by a larger program that is running in a computer. Some DLLs referred to as device drivers enable the computer to communicate with a specific type of hardware, such as a printer or scanner.

Because a DLL file is not loaded into random access memory (RAM) together with the main program, its use saves space in RAM. DLL files have the file name suffix .dll and are dynamically linked with the program that uses them during program execution. Only when a DLL file is needed is it loaded into RAM and run.

Helpers

Collectively plug-ins, ActiveX, and Java applets are referred to as helpers. They are given this name because they handle documents on the client browser that the browser by itself is not capable of doing. Figure 2.3 provides a general overview of browser components. Note that while HTML decoding is built into each browser, other components are optional.

Plug-ins

A plug-in is a computer program that functions as an extension of a browser, adding some specific functionality such as displaying multimedia content inside a Web page. Some examples of plug-ins include Shockwave, RealPlayer, and QuickTime for multimedia, Net Zip, and Neptune. NetZip permits compression of data while Neptune supports ActiveX for Netscape so it operates similar to Microsoft's Internet Explorer.

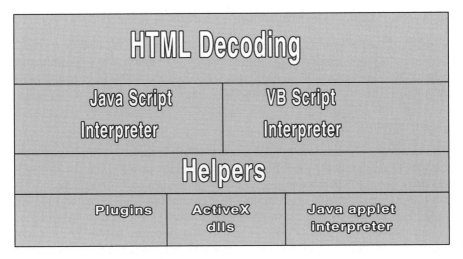

Figure 2.3 Browser components.

Originally, Microsoft went to great lengths to make Internet Explorer compatible with Netscape. From JavaScript to HTML to plug-ins, Internet Explorer functioned similar to Netscape. In fact, early versions of Internet Explorer could even read the Netscape plug-in directory. Unfortunately, as Internet Explorer gained market share its compatibility with Netscape diminished.

Java

Java is a high-level programming language that is unusual because it is both compiled and interpreted. Using a compiler, a Java program is first translated into an intermediate language called Java bytecodes, which represents platform independent coding that is subsequently interpreted on the Java platform. The interpreter parses or divides it into small components so each Java bytecode instruction is executed on the computer. Although Java program compilation only occurs once, each time the program is executed it is interpreted.

Java bytecodes

Java bytecodes can be viewed as machine code instructions for what is referred to as a Java Virtual Machine (Java VM). Every Java interpreter, including a Web browser that can run applets and a specialized development tool, can be considered as an implementation of a Java VM. Java bytecodes permit a programmer to develop software that can operate on any Java VM, permitting platform independence as long as the platform includes a Java interpreter. Now that we have an appreciation for the Java VM, let's discuss another feature of Java that provides additional functionality. That feature is the Java Application Programming Interface (API).

The Java Application Programming Interface (Java API)

The Java Application Programming Interface (Java API) represents a collection of predefined software components, which when invoked performs predefined functions. The Java API is grouped into libraries of related classes and interfaces, with each library referred to as a package.

Java programs

The most common types of Java programs are applets and applications. Of the two, most readers are probably more familiar with Java applets as they represent a program that executes within a Java-enabled browser. An applet represents a program written in Java that can be included within an HTML Web page. When you use a Java-enabled browser to view a page that contains a Java applet, the applet's code is transferred from the server to the client where it is executed by the browser's Java VM. In comparison, an application represents a standalone program that executes directly on a Java enabled computer platform.

Several varieties of Java applications warrant a brief discussion. First, a special type of Java application known as a server supports clients on a network. Examples of Java servers include Web servers, print servers, proxy servers, and other types of servers that transmit Java applications. Another type of Java program is a servlet, which can be viewed as an applet that runs on a server. Servlets are commonly used in constructing interactive Web applications in place of CGI scripts.

Another type of Java program that deserves mention is JavaScript. JavaScript represents a cross-platform object-based scripting language. JavaScript is a small, concise language designed to be embedded in other products and applications including Web browsers. Inside a browser, JavaScript can be connected to the objects of its environment in effect providing program control over its environment.

JavaScript contains a core set of objects, such as Array, Date, Math, a core set of operators, control structures, and statements referred to as language elements.

This core JavaScript can be extended for client-side and server-side operations. In client-side core, JavaScript is extended by supplying objects that provide a degree of control over a browser, such as letting a user place elements into a form or navigate through a Web page. In comparison, server-side Java extends the core scripting language by providing the elements necessary to run JavaScript on a server. One example of a server-side JavaScript extension would be an application that communicates with a back-end database.

The integration of a JavaScript program into a Web page is obtained using the HTML <SCRIPT> tag. The following HTML coding indicates an example of the integration of a JavaScript program into an HTML Web page.

```
<HTML>

<HEAD>

<TITLE> JavaScript Example </TITLE>

<SCRIPT LANGUAGE = "JavaScript">

MsgBox "Welcome to popcorn.com."

</SCRIPT>
```

Visual Basic Scripting (VBScript)

Microsoft's Visual Basic Scripting (VBScript) can be viewed as a more powerful and potentially more dangerous extension to HTML than Java applets.
A VBScript communicates with host applications using Windows Script.
Using Windows Script, Microsoft's Internet Explorer, and other host applications, do not require special coding for each scripting component enabling
a computer to compile scripts, obtain and call entry points, and even create a
standard language run time for scripting. In a client-server environment, a
VBScript enabled browser will receive scripts imbedded within a Web page.
The browser will parse and process the script.

Similar to including JavaScript into a Web page, integrating a VBScript
is obtained using the HTML <SCRIPT> tag. The following HTML code
indicates an example of integrating a VBScript into an HTML Web page.

```
<HTML>

<HEAD>

<TITLE> VBScript Example </TITLE>

<SCRIPT LANGUAGE = "VBScript">

MsgBox "Welcome to popcorn.com."

</SCRIPT>
```

Using the SCRIPT LANGUAGE tag, the Web browser is informed how
to interpret the code. Although Netscape browsers and some versions of
Microsoft's Internet Explorer support JavaScript, only Microsoft's Internet
Explorer supports VBScript.

If a browser does not support a particular scripting language embedded
into a Web page, it will either display the script as part of the Web page or
hide the script from view. Hiding the script occurs when the script is encased
in comment tags (<! - - and - - >) and the browser simply ignores the script.
For example, returning to our prior script example, if we encase the script as
follows, it will be ignored by a browser that does not support VBScript.

```
<HTML>

<HEAD>
```

```
<TITLE> VBScript Example </TITLE>

<SCRIPT LANGUAGE = "VBScript">

<!- -

MsgBox "Welcome to PopCorn.ccom!"

-- >

</SCRIPT>

</HEAD>
```

ActiveX

ActiveX represents a set of rules that defines how applications should share information. Developed by Microsoft, ActiveX has its roots in two Microsoft technologies: Object Linking and Embedding (OLE) and Component Object Model (COM).

Programmers can develop ActiveX controls in a variety of programming languages, such as C, C++, Java, and Visual Basic. An ActiveX control is similar to a Java applet. However, unlike a Java applet that is limited in capability and cannot perform disk operations, ActiveX controls have full access to the Windows operating system. While this capability provides ActiveX controls with more capability than Java applets it also entails a security risk. To control this risk, Microsoft developed a registration system that enables browsers to identify and authenticate an ActiveX control prior to its download.

2.3 Server Operations

In the client-server model illustrated in Figure 2.1, we noted that the server layer can consist of one or more computers. While you can always expect a Web server in the client-server model, it's also possible that the Web server will communicate with an application server. The application server in turn can communicate with a database that resides either on the application server or on a separate device, such as a back-end database or a Redundant Array of Inexpensive Disks (RAID) array. In this section, we will briefly examine Web server operations including the way they interconnect to application servers and a database.

Evolution

The modern Web server can be viewed as a descendent of Web servers developed at the U.S. National Center for Supercomputing Applications (NCSA) where Mosaic, the browser that evolved into Netscape, was developed and CERN, the European Organization for Nuclear Research. As work on client browser and Web server software progressed during the late 1990s

Table 2.7 Common Web Server Application Programs

Program	Operating Environment
Internet Information Server	Microsoft bundles this product with its Windows 2000/2003 Server
Apache	An open source, no-charge-for-use HTTP server that operates under UNIX, Linux and Windows
SunONE (renamed Java System)	Originally developed by Netscape jointly with Sun Microsystems, versions operate under Windows and UNIX
WebSTAR	A server suite from 4D, Inc. that operates on a Macintosh platform
Red Hat Content Accelerator	A kernel-based Web server that is limited to supporting static Web pages in a Linux environment

at NCSA and CERN, commercial applications on the Internet rapidly expanded resulting in software developers offering products for this rapidly expanding market. Although there are a limited number of Web server programs to choose from today, each operates in a similar manner. All communications between Web clients and the server use HTTP, and the Web server commences operation by informing the operating system it operates under, that it's ready to accept communications through a specific port. That port is 80 for HTTP, which is transported via TCP, and 443 when secure HTTP (https) is employed.

Common Web server programs

A number of Web server application programs operate under different operating systems. Some Web server programs are bundled with general server software designed to operate under several versions of Windows on Intel Pentium and AMD platforms. Other Web server programs operate under UNIX and Linux operating systems. Table 2.7 lists some of the more popular Web server programs and a brief description of the operating environment of the program.

Web server directories

In general, Web servers can be considered to have two separate types of directories. One directory is created by the operating system and commences at the beginning of the disk drive, which is considered to represent the root directory. The second directory commences from a location on the disk drive under which Web documents are normally stored. This Web directory also has a root, referred to as either the document root or the Home Directory local path.

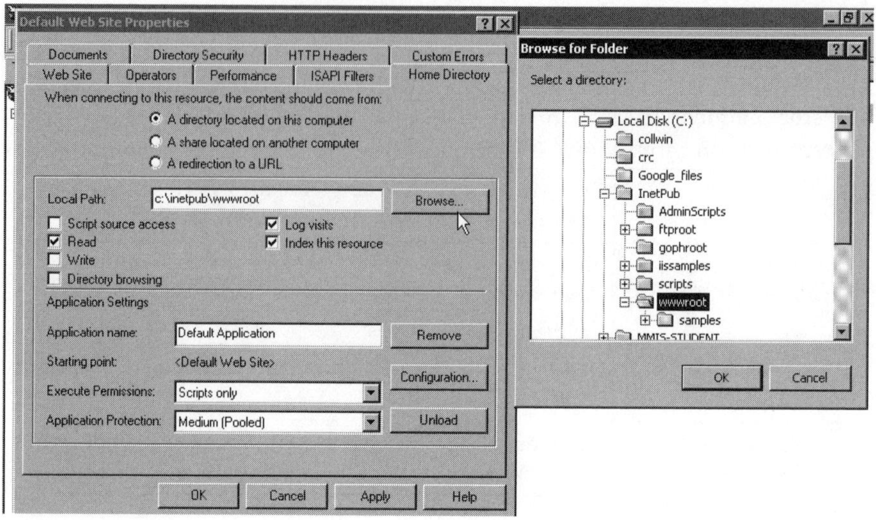

Figure 2.4 The root or home directory is specified with respect to the disk root.

The left portion of Figure 2.4 illustrates the Microsoft Internet Information Server's Default Web Site Properties dialog box with its Home Directory tab selected. Note that the local path is shown as c:\inetpub\wwwroot, which represents the default home directory from which server content flows in response to a client query. By clicking on the Browse button, the Browse for Folder window is opened, which is shown in the right portion of Figure 2.4.

Let's assume the computer URL or site name is www.popcorn.com and you store a document named welcome in the wwwroot directory. Then, c:\inetpub\wwwroot\welcome represents the welcome document directory address. The URL request of http://www.popcorn.com would elicit the return of the Welcome Web page from the path c:\inetpub\wwwroot\welcome.html. From this we can note that a Web server maps URL requests to a directory structure on the computer in terms of the Web home page or document root. In addition to being able to specify a local path or document root, the server operator can configure a Web server to generate content from a share on another computer or a redirection to a URL. The redirection permits requests to be redirected from one directory to another directory that could be on the same computer or on a different computer located thousands of miles away.

Server characteristics

A modern Web server has several characteristics that deserve mentioning. Among those characteristics are the ability to support multiple sites on the same computer and the capability to provide clients with documents from the document roots or Internet directories on other servers. Supporting multiple sites is referred to as virtual hosting while the ability to provide documents from other servers turns the Web server into a proxy server. In addition to supporting Web page delivery, modern Web servers include support for File Transfer Protocol (FTP), Gopher, News, email, and database access. Web servers also commonly run CGI scripts and servlets, the servlets representing a compiled Java class. Because CGI scripts and servlets can significantly enhance client-server operations, let's focus on both.

Common Gateway Interface (CGI scripts)

The Common Gateway Interface (CGI) represents a standard for interfacing external applications with servers. To understand the need for CGI, you need to realize that a plain HTML coded document is static and does not change. This means if you need to vary the document, such as placing an order or entering a request for an item that could range from the price of a stock to the weather in a ZIP code, you need a tool to output dynamic information. A CGI program provides that tool because it is executed in real-time, which enables it to output dynamic information.

CGI dates to the early development of the Web when companies wanted to connect various databases to their Web servers. The connection process required a program that, when executed on a Web server, would transmit applicable information to the database program, receive the results of the database query, and return the results to the client. Because the connection process between the server and the database functioned as a gateway, the resulting standard acquired the name Common Gateway Interface.

A CGI program can be written in any language that can be executed on a particular computer. C, C++, Fortran, PERL, and Visual Basic represent some of the languages that can be used in a CGI program. In a UNIX environment, CGI programs are stored in the directory /cgi-bin. In a Microsoft Internet Information Server environment, CGI programs are stored in c:\internetpub\wwwroot\scripts. That location holds CGI programs developed using a scripting language, such as PERL.

One of the most common uses for CGI scripts is for form submission, with a CGI script executing on the server processing the entries in the form transmitted by the client. CGI scripts can be invoked by specifying their URL directly in HTML or embedding them in another scripting language, such as JavaScript.

Servlets

A servlet represents a compiled Java class. Similar to CGI, servlets operate on a server and are called through HTML. When a Web server receives a request for a servlet, the request is passed to the servlet container. The container then loads the servlet. Once the servlet is completed, the container reinitializes itself and returns control to the server.

Although similar to CGI, servlets can be faster because they run as a server process and have direct access to Java APIs. Because servlets are written in Java, they are platform independent.

In concluding our discussion of servlets, let's turn to a related cousin; Java Server Pages (JSP). JSPs are similar to servlets in that they provide processing and dynamic content to HTML documents. JSPs are normally used as a server-side scripting language and are translated by the JSP contained into servlets.

Prior to moving on to the application server, a few words are in order concerning Microsoft server standard referred to as Internet Server Application Programming Interfaces (ISAPI).

Internet Server Application Programming Interfaces (ISAPI)

ISAPI represents a Microsoft server specific standard to load a DLL into the address space of a server to interpret a script. Although similar to CGI, the ISAPI is faster because there is no need for a server to spawn a new executable, as is the case when a CGI script is used. Instead, a DLL is loaded, which provides the interpretation of the script.

Application servers

If we briefly return to Figure 2.1, we can note that the application server represents a tier 2 device along with a conventional Web server. Both a Web server and application server can reside on the same or on separate computers. The application server commonly referred to as an appserver, can range in scope from a program that handles application operations between a client and an organization's back-end databases, to a computer that runs certain software applications. In this section, we will focus on several application server models and various popular software products that can be used between the application server and the tier 3 database layer.

Access

For many clients access to an application server is transparent. The client will use a browser to access a Web server. The Web server, depending on its coding, can provide several different ways to forward a request to an application server. Some of those ways include using CGI, Microsoft's Active Server Page, and the Java Server Page.

An Active Server Page (ASP) is an HTML page that includes one or more scripts processed by the Web server prior to the Web page being transmitted to the user. An ASP file is created by including a VBScript or JavaScript in an HTML file or by using ActiveX Data Objects program statements in a file. In comparison, a JSP uses servlets in a Web page to control the execution of a Java program on the server.

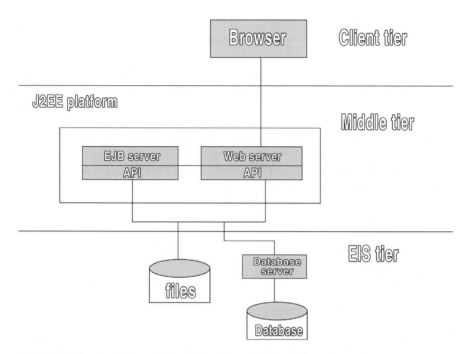

Figure 2.5 The three-tier relationship when a J2EE platform is used.

Java application servers

Java application servers are based on the Java 2 Platform, Enterprise Edition (J2EE). J2EE employs a multi-tier model that normally includes a client tier, middle tier, and an Enterprise Information Systems (EIS) tier, with the EIS tier storing applications, files, and databases. In comparison, the middle tier can consist of a Web server and an Enterprise Java Bean (EJB) server.

Using J2EE requires an accessible database. When using J2EE you can access a database using Java Database Connectivity (JDBC), Application Program Interfaces (API), System Query Language-Java (SQLJ), or Java Data Objects (JDO).

JDBC represents an industry standard API for database-independent connectivity between the Java programming language and a wide range of databases. SQLJ represents a specification for using SQL with Java, while JDO represents an API standard Java model that makes Plain Old Java Objects (POJOs) persistent in any tier of an enterprise architecture. Figure 2.5 illustrates the relationship of the three tiers to one another when a J2EE platform is employed. In examining Figure 2.5, note that EJB represents a server-side component architecture for the Java 2 platform. EJB enables development of distributed, secure, and portable applications based on Java.

There are two types of EJBs — entity beans and session beans — with the entity beans representing an object with special properties. When a Java program terminates, any standard objects created by the program are lost including any session beans. In comparison, an entity bean remains until it is explicitly deleted. Any program on a network can use an entity bean as long as the program can locate it. Because it's easy to locate data on permanent storage, entity beans are commonly stored within a database. This allows the entity bean to run on a server machine. When a program calls an entity bean, control is passed to the server and the program thread stops executing. On completion, control is restored to the calling program and the program resumes execution.

General server tools

In concluding our discussion concerning application servers, we need to consider two general server tools — Cold Fusion and Open Database Connectivity (ODBC). Cold Fusion is a popular development toolset from Macromedia that enables databases to be integrated with Web pages. Cold Fusion Web pages include tags written in Cold Fusion Markup Language (CFML) that simplifies the integration with databases, eliminating the necessity to use more complex languages, such as C++. For example, using Cold Fusion you could create a Web page that asks a user for their sex and age, information that would be used by a Web server to query a database for an insurance premium that would be presented in HTML code for display on the user's browser.

The second tool, ODBC, represents a standard database access method developed by the SQL Access group. By inserting a middle layer, called a database driver, between an application and the database management system (DBMS), it is possible to access data from any application regardless of which DBMS is handling the data. The database driver translates application data queries into commands that the DBMS understands, requiring that the application and the DBMS are ODBC compliant. Now that we have an appreciation for two popular tools used in client-server operations, we will conclude this chapter by examining the efficiency of the architecture as the distance between client and server layers increases.

2.4   *Distance Relationship*

In the three-tier client server model, the first two tiers are usually separated from one another while the third tier is normally located with the second tier. When examining the distance relationship between tiers we can view the three-tier model as being a two-tier one. We can examine the effect on client-server architecture as the distance between client and server increases. We can use two popular TCP/IP tools; the ping and the traceroot programs contained in modern operating systems.

```
Command Prompt                                                    _ □ ×
Microsoft Windows 2000 [Version 5.00.2195]
(C) Copyright 1985-2000 Microsoft Corp.

C:\WINNT\Profiles\Administrator>ping yahoo.com

Pinging yahoo.com [216.109.112.135] with 32 bytes of data:

Reply from 216.109.112.135: bytes=32 time=20ms TTL=57
Reply from 216.109.112.135: bytes=32 time=20ms TTL=57
Reply from 216.109.112.135: bytes=32 time=11ms TTL=57
Reply from 216.109.112.135: bytes=32 time=20ms TTL=57

Ping statistics for 216.109.112.135:
    Packets: Sent = 4, Received = 4, Lost = 0 (0% loss),
Approximate round trip times in milli-seconds:
    Minimum = 11ms, Maximum =  20ms, Average =  17ms

C:\WINNT\Profiles\Administrator>
```

Figure 2.6 Using ping to ascertain the round-trip delay to Yahoo.com.

Using ping

The ping utility program by default generates four data packets that are transmitted to a defined destination. That destination can be specified either as a host address or as a dotted decimal address. The recipient of the packets echoes the packets back to the originator. Because the originator knows when it transmitted each packet, it is possible to compute the time in terms of a round trip delay.

Technically, ping is based on the Internet Control Message Protocol (ICMP). ICMP type 8 (Echo) messages are transmitted to an indicated destination, which responds with ICMP type 0 (Echo Reply) messages.

To illustrate the effect of distance on client-server operations, let's examine the round trip delays between this author's computer located in Macon, Georgia, and two servers, one located in the United States and the second located in Israel. Figure 2.6 illustrates the pinging of the Yahoo! server. In examining the entries in the referenced illustration, you will note that four replies are listed from the IP address 216.109.112.135, which represents the IP address of the Yahoo! server. The round trip delay times are indicated by time = and have values of 20, 20, 11, and 20 milliseconds (ms), respectively. Below the last reply line, the program generates a summary of statistics for pinging the destination. In the example, shown for pinging Yahoo!, the average round trip delay was indicated to be 17 ms.

For our second example, this author pinged the server Logtel.com, a Data Communications seminar organization located in Israel. Figure 2.7 illustrates the results associated with pinging that location. As you might expect, the round trip delay has considerably increased because of the increased distance between the client and the server. However, what may

```
[□] Command Prompt                                                        _ □ ×
Microsoft Windows 2000 [Version 5.00.2195]
(C) Copyright 1985-2000 Microsoft Corp.

C:\WINNT\Profiles\Administrator>tracert www.yahoo.com

Tracing route to www.yahoo.akadns.net [216.109.118.71]
over a maximum of 30 hops:

  1   <10 ms   <10 ms   <10 ms  205.131.176.1
  2   <10 ms    10 ms   <10 ms  s4-0-8.hsa1.atl1.bbnplanet.net [4.24.209.73]
  3   <10 ms    10 ms   <10 ms  ge-6-0-0.bbr1.atlanta1.level3.net [64.159.1.245]
  4    20 ms    20 ms    10 ms  as-2-0.bbr1.washington1.level3.net [64.159.1.2]
  5    20 ms    20 ms    10 ms  ge-1-1-53.car1.washington1.level3.net [4.68.121.69]
  6    20 ms    20 ms    10 ms  4.79.228.6
  7    20 ms    20 ms    10 ms  v132.bas1-m.dcn.yahoo.com [216.109.120.150]
  8    20 ms    20 ms    20 ms  p8.www.dcn.yahoo.com [216.109.118.71]

Trace complete.

C:\WINNT\Profiles\Administrator>_
```

Figure 2.7 Using ping to determine the round trip delay to a server located in Israel.

not be apparent is the reason for the increase, which is approximately an order of magnitude, from an average of 17 ms for pinging Yahoo! to an average of 223 ms when pinging the Logtel server. To obtain an appreciation of the key reason behind the increased round trip delay we need to turn to the second TCP/IP tool, traceroot.

Using traceroot

Traceroot is a program that traces the route from source to destination. Under Microsoft Windows, the name of the program was truncated to tracert.

Tracert invokes a series of ICMP Echo messages that vary the time-to-live (TTL) field in the IP header. The first IP datagram that transports the ping has a TTL field value of one. Thus, when the datagram reaches the first router along the path to the destination, the router decrements the TTL field value by one and since the result is zero, throws the datagram into the great bit bucket in the sky and returns an ICMP message Type 11 (Time Exceeded) to the originator. The ICMP message returned to the sender includes the router's IP address and may include information about the router. The originator increments the TTL value by one, and retransmits the ping, allowing it to flow through the first router on the path to the destination. The second router then returns an ICMP Type 11 message and the process continues until the tracert program's ping reaches the destination or the default number of hops used by the program is reached.

Figure 2.8 illustrates the Microsoft tracert program to trace the path from the author's computer in Macon, Georgia, to the Yahoo! server. In comparison, Figure 2.9 illustrates the tracert program to trace the path to the Logtel server located in Israel.

In comparing the two uses of the tracert program, we can note that there are eight router hops to Yahoo.com while there are 17 hops to the Logtel server. Because each router hop requires some processing time, we can

```
Command Prompt                                                    _ □ ×

C:\WINNT\Profiles\Administrator>ping logtel.com

Pinging logtel.com [212.150.150.132] with 32 bytes of data:

Reply from 212.150.150.132: bytes=32 time=220ms TTL=114
Reply from 212.150.150.132: bytes=32 time=221ms TTL=114
Reply from 212.150.150.132: bytes=32 time=220ms TTL=114
Reply from 212.150.150.132: bytes=32 time=231ms TTL=114

Ping statistics for 212.150.150.132:
    Packets: Sent = 4, Received = 4, Lost = 0 (0% loss),
Approximate round trip times in milli-seconds:
    Minimum = 220ms, Maximum =  231ms, Average =   223ms

C:\WINNT\Profiles\Administrator>
```

Figure 2.8 Using Tracert to observe the path from the author's computer to the Yahoo! server.

```
Command Prompt                                                    _ □ ×

Tracing route to www.logtel.com [212.150.150.132]
over a maximum of 30 hops:

  1   <10 ms   <10 ms   <10 ms  205.131.176.1
  2   <10 ms   <10 ms    10 ms  s4-0-8.hsa1.atl1.bbnplanet.net [4.24.209.73]
  3    10 ms   <10 ms   <10 ms  ge-6-1-0.bbr2.atlanta1.level3.net [64.159.3.17]
  4   <10 ms   <10 ms    10 ms  so-5-0-0.gar1.atlanta1.level3.net [4.68.96.22]
  5   <10 ms    10 ms   <10 ms  uunet-level3-oc12.atlanta1.level3.net [4.68.127.58]
  6    10 ms   <10 ms    10 ms  0.so-2-1-0.xl2.atl5.alter.net [152.63.84.154]
  7   <10 ms    10 ms    10 ms  0.so-0-0-0.tl2.atl5.alter.net [152.63.10.106]
  8    20 ms    31 ms    20 ms  0.so-6-0-0.tl2.nyc9.alter.net [152.63.13.10]
  9    30 ms    20 ms    20 ms  0.so-1-2-0.xl2.nyc4.alter.net [152.63.21.13]
 10    20 ms    20 ms    20 ms  pos7-0.ig5.nyc4.alter.net [152.63.35.37]
 11    90 ms   100 ms    90 ms  Barak-gw6.customer.alter.net [157.130.255.42]
 12   220 ms   211 ms   210 ms  bb3-2.ser1-0-0.barak.net.il [212.150.232.66]
 13   210 ms   210 ms   221 ms  212.150.234.78
 14   210 ms   210 ms   211 ms  212.29.206.198
 15   230 ms   210 ms   211 ms  barak-1-acc-3.barak.net.il [206.49.94.116]
 16   220 ms   220 ms   211 ms  62.90.17.98
 17      *              221 ms   220 ms  webnt1.barak.net.il [212.150.150.132]

Trace complete.

C:\WINNT\Profiles\Administrator>
```

Figure 2.9 Using Tracert to examine router hop delays to a server located in Israel.

attribute a portion of the delay to the number of router hops traversed from the source to the destination. However, as the radio commentator Paul Harvey is fond of saying, "that's only part of the story."

If you carefully examine the tracert to Logtel.com shown in Figure 2.9, you will note that for the first 10 hops the delay was under 20 ms. It wasn't until hop 11 that the delay appreciably increased, going from 20 to either 90 or 100 ms, depending on which of the three traces occurred. At hop 10, the router description indicates it was in New York City, while at hop 11, the router appears to be a gateway that funnels traffic to Israel. At hop 12, the router description indicates it is in Israel. Approximately 70 to 80 ms of delay can be attributed to the router acting as a funnel for traffic to Israel at

hop 11, while another 110 to 120 ms of delay can be attributed to the propagation delay between the New York gateway and Israel. Once the packets
arrive in Israel, the delay from hop 12 to hop 17 is approximately 10 ms.
Thus, the primary delays are the gateway funneling traffic from the East
Coast to Israel and propagation delays.

If your organization was located in the United States and had customers
in Israel, the results of the tracert would be reversed. Israeli users would
experience bottlenecks due to propagation delays and traffic being funneled
through a peering point located in New York City. We can expand this
situation to users in Western Europe, South America, Japan, China, and other
locations around the globe that, when accessing servers located in the United
States, would also experience bottlenecks from traffic flowing through peering points and propagation delays. To alleviate these delays your organization can move servers storing duplicate information closer to the ultimate
user, which is the key rationale for content delivery networking. However,
because only the largest companies may be able to afford placing servers at
distributed locations around the globe, most content delivery methods
depend on using a third party that provides the service under contract to
customers.

chapter three

Understanding TCP/IP

The ability to understand technical details associated with content delivery requires an understanding of the Transmission Control Protocol Internet Protocol (TCP/IP) protocol suite. In this chapter, we will briefly review the TCP/IP protocol suite, focusing on the fields that govern the identification of applications and the delivery of IP datagrams.

3.1 The Transmission Control Protocol Internet Protocol (TCP/IP) Suite

The TCP/IP protocol suite dates to the work of the Advanced Research Projects Agency (ARPA) during the 1970s and early 1980s. During that time, the quest to interconnect computers resulted in the development of a series of protocols that evolved into the modern TCP/IP protocol suite.

Protocol suite components

Figure 3.1 illustrates the major components of the TCP/IP protocol suite and their relationship to the International Standards Organization (ISO) Open System Interconnection (OSI) Reference Model. In examining Figure 3.1, note that the TCP/IP protocol suite does not specify a physical layer nor does it specify a data link layer. Instead, the protocol suite uses its Address Resolution Protocol (ARP) as a way to enable the protocol suite to operate above any data link layer that is capable of transporting and responding to ARP messages. This enables the TCP/IP protocol suite to interoperate with Ethernet, Fast Ethernet, Gigabit Ethernet, and Token-Ring local area networks.

In examining the relationship of the TCP/IP protocol suite to the OSI Reference Model shown in Figure 3.1, several additional items warrant mention. First, although applications are transported at layer 5 in the protocol suite, they correspond to layers 5, 6, and 7 of the OSI Reference Model. Secondly, one of two protocols, TCP or User Datagram Protocol (UDP), commonly carries applications. As we will note later in this chapter, TCP

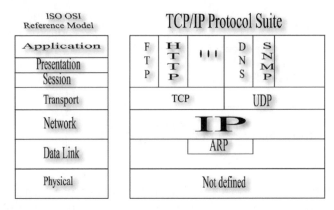

Figure 3.1 Major components of the TCP/IP Protocol suite.

provides a connection oriented, reliable transport facility while UDP provides a best-effort, non-reliable transport facility.

Applications transported by TCP and UDP are identified by Destination Port fields within each Transport Layer header. When TCP is used, the TCP header plus application data is referred to as a TCP segment. In comparison, when UDP is used as the Transport Layer, the UDP header plus application data is referred to as a UDP datagram. An IP datagram is formed by prefixing an IP header to either a TCP segment or UDP datagram. Because the IP header contains Source and Destination address fields, routing occurs by examining the IP header fields. Since different applications are identified using port numbers within the TCP or UDP header, it is possible for a common destination, such as a corporate server, to provide Web, email, and file transfer support. As we probe deeper into the protocol suite using IP addresses, TCP and UDP port numbers to define applications destined to specific devices will become clearer.

Physical and data link layers

As a review, the physical layer represents the electrical and mechanical components necessary to connect to the network. In comparison, the data link layer uses a protocol to group information into data packets that flow on the network. Because the data link uses a layer 2 protocol, source and destination addresses are indicated in terms of Media Access Control (MAC) addresses.

Media Access Control (MAC) addressing

MAC addresses are 48 bits or 6 bytes in length, subdivided into a vendor code and identifier that corresponds to the vendor code. The Institute of Electrical & Electronics Engineers (IEEE) assigns vendor codes and the vendor or manufacturer burns into each Local Area Network (LAN) adapter's

Read Only Memory (ROM), a unique 48-bit address using the assigned vendor code but varying the identifier number in each adapter. If the vendor is successful at marketing their LAN adapters, they will request additional vendor codes from the IEEE and repeat the previously described process.

Although using MAC addresses ensures there will be no duplicate addresses on a LAN, the absence of any network identification made it difficult to interconnect LANs. Without a LAN identifier, it is difficult to note the destination LAN for a layer 2 frame. In fact, the method of routing data between LANs was originally based on bridging; a layer 2 technique in which the 48-bit MAC addresses were used to determine if a frame should flow across a bridge. Because early bridges interconnected LANs located in close proximity to one another, it was difficult to interconnect LANs located in different cities or even at different locations within the same city. To overcome these limitations, network protocols, including TCP/IP that oper-ate at the network layer, have the ability to assign unique network addresses to each network, making it possible to route data between networks based on a destination network address contained in each packet. Prior to discuss-ing the network layer in the TCP/IP protocol suite, we need to cover ARP, which functions as a bridge between the network layer and the data link layer as shown in Figure 3.1. However, prior to discussing ARP a few words are in order concerning layer 3 addressing in the TCP/IP protocol suite.

Layer 3 addressing

There are two versions of the IP in use: Internet Protocol Version 4 (IPv4) and Internet Protocol Version 6 (IPv6). IPv4 uses 32-bit addressing while IPv6 uses 128-bit addressing. Because approximately 99 percent of organiza-tions currently use IPv4, we will focus in this section on the 32-bit addressing scheme used by IPv4.

Under IPv4 there are five address classes, referred to as Class A through Class E. The first three address classes are subdivided into network and host portions as indicated in Figure 3.2.

Class A addresses were assigned to very large organizations. If you examine Figure 3.2, you will note that one byte of the four 8-bit bytes in the address is used to denote the network while the remaining three bytes in the 32-bit address are used to identify the host on the network. Although an 8-bit byte normally provides 256 unique addresses, under IPv4 the first bit in the address field is set to a binary one to identify a Class A address, reducing the number of unique bits in the address to seven. There can only be a maximum of 2^7 or 128 Class A addresses. Since one Class A address represents a loopback address, while the IP address of 0.0.0.0 is used for the default network, this reduces the number of available Class A addresses to 126 and explains why many years ago all Class A addresses were assigned. Because three bytes of the Class A network address are used to identify each host, each Class A network can support $2^{24} - 2$ or 16,777,214 distinct hosts. We subtract two from the total because a host address of all zeros is used to

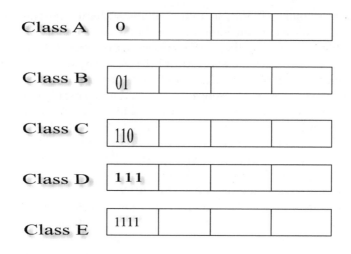

Figure 3.2 IP address classes.

identify the network (this network) while a host address of all ones represents the network broadcast address.

Returning to Figure 3.2, we can note that a Class B address extends the number of bytes used to identify the network to two, resulting in two bytes being used to identify the host on the specified network. Because the first two bits in the 32-bit address are used to identify the address as a Class B address, 14 bits are available to identify 2^{14} (16,384) unique networks. There are considerably more Class B addresses available for use than Class A addresses. Because each Class B network uses two bytes to define the host address on the network, there are $2^{16} - 2$ or 65,534 possible hosts on each Class B network. Class B addresses were typically issued to large organizations.

The third IPv4 address that is subdivided into network host portions is the Class C address. A Class C address uses three bytes to define the network portion of the address while the remaining byte in the 32-bit address is used to identify the host on the network. A Class C address also uses the first three bits in the first byte in the address to identify the address as a Class C address. There are 2^{21} (2,097,152) unique Class C network addresses.

A Class C network address is the most popularly used IP address, commonly issued to small to mid-sized organizations. Because only one byte is available to define the hosts on a network, a Class C address supports the least number of network hosts. The number of unique hosts on a Class C network is limited to $2^8 - 2$ or 254. Two is subtracted from the computation because similar to Class A and Class B addresses, two Class C addresses have special meanings and are not used to identify a specific host on a network. Those addresses are 0 and 255. A host address of zero is used to indicate this network while a host address of 255 represents the broadcast address of the network. Thus, the inability to use those two addresses as host addresses results in 254 hosts being capable of having unique addresses on a Class C network.

Although Class A, B, and C are the most commonly used IPv4 addresses, two additional Class addresses warrant a brief mention. Those addresses are Class D and Class E addresses. Class D addresses are used for multicast operations while Class E addresses are reserved for experimentation. Now we have an appreciation for the five types of IPv4 addresses, let's turn to the way network addresses are translated into MAC addresses. That translation, as we previously noted is accomplished using the ARP.

Address Resolution Protocol (ARP)

When a router receives an IPv4 packet addressed to a specific network, the destination address is specified as a 32-bit address. However, data delivery on the LAN is based on layer 2 MAC addresses. This means the 32-bit layer 3 IP address received by a router must be translated into a 48-bit MAC address for the packet to be delivered by the layer 2 protocol.

When a router receives a layer 3 packet, it first checks its cache memory to determine if a previous address translation occurred. If so, it forms a layer 2 frame to deliver the layer 3 packet using the previously learned layer 2 MAC address as the destination address in the frame. If no previous address translation occurred, the router will use ARP as a way to determine the layer 2 address associated with the layer 3 destination address. The router forms an ARP packet, indicating the IP address it needs to learn. The ARP packet is transported as a layer 2 broadcast to all hosts on the network. The host that is configured with the indicated IP address responds to the broadcast with its MAC address. The router uses ARP to learn the MAC address required to deliver layer 3 addressed packets to their correct destination on a layer 2 network where data delivery occurs using MAC addresses. Now we have an appreciation for using ARP to enable layer 3 addressed packets to be delivered on layer 2 networks, we can turn to the higher layers of the TCP/IP protocol suite.

The network layer

The IP represents a network layer protocol that enables IP datagrams to be routed between source and destination networks.

Figure 3.3 illustrates the formation of an IP datagram, showing the relationship of the IP header to the two transport layer headers commonly used in the TCP/IP protocol suite — TCP and UDP. Note the application data is first prefixed with either a TCP or an UDP header prior to being prefixed with an IP header.

The IP header consists of 20 bytes of information subdivided into specific fields. An option exists within the IP header that enables the header to be extended by adding optional bytes; however, this extension is rarely used.

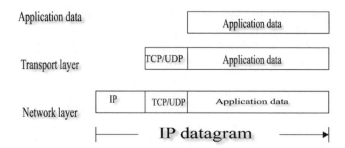

Figure 3.3 Forming an IP datagram.

Internet Protocol (IP) header

The top portion of Figure 3.4 illustrates the fields within the IPv4 header. The lower portion provides a brief description of each of the IPv4 header fields. For the purpose of content delivery, the key fields of interest are the time-to-live (TTL), the Protocol, the 32-bit source IP address, and 32-bit destination IP address fields.

Time-to-Live (TTL) field

The TTL field indicates the number of hops the IP packet can traverse before it is discarded. The purpose of this field is to ensure that packets do not continuously flow through the Internet if the destination is not located. To prevent endless wandering, routers decrement the TTL field value and, if the result is zero, discard the packet. In a content delivery networking environment, the movement of Web server data closer to the requestor commonly ensures that the decrement of the TTL field value never reaches zero, which would require a packet to be sent to the great bit bucket in the sky.

Protocol field

The Protocol field is 8-bits in length and indicates the type of transport packet carried in the IP datagram. Because the Protocol field is 8-bits in length, up to 256 protocols can be defined. Some of the more popular protocols are the Internet Control Message Protocol (decimal one), the TCP, which is defined by decimal six in the IP Protocol field, and the UDP, which is defined by decimal 17 in the IPv4 Header's Protocol field. The IPv4 Protocol field value defines the upper layer protocol used to transport data. Later in this chapter, we will note that both TCP and UDP headers include Port fields whose values define the application carried at the transport layer.

Source and destination addresses fields

The remaining two fields of interest in the IPv4 header, with respect to content delivery, are the Source and Destination IP addresses. Each address is 32-bits

Ver	IHL	ToS	Size
Identification		FLAGS	Fragment offset
TTL		Protocol	Checksum
32-bit Source Address			
32-bit Destination Address			
Options (if any)			

Legend:

Ver	Version number
IHL	IP Header Length (number of 32-bit words in the header)
ToS	Type of Service byte, now known as the Differentiated Services Code Point (DSCP)
Size	Size of the datagram in bytes (header plus data)
Identification	16-bit number which together with the source address uniquely identifies the packet
FLAGS	Used to control if a router can fragment a packet
Fragment offset	A byte count from the start of the original packet set by the router that performs fragmentation.
TTL	Time to Live or the number hops a packet can be routed over
Protocol	Indicates the type of packet carried
Checksum	Used to detect errors in the header
Source address	The IP address of the packet originator
Destination address	The IP address of the final destination of the packet

Figure 3.4 The IPv4 header.

in length, with the Source Address indicating the originator of the packet while the Destination address indicates the ultimate recipient of the packet.

As explained earlier in this chapter, under IPv4 there are five address Classes, labeled A through E. The vast majority of data traffic on the Internet occurs using address Classes A, B, and C, while address Class D is used for multicast transmission and address Class E is reserved for experimental operations. Concerning address Classes A, B, and C, each of those addresses are subdivided into a network and host portion. Using those addresses conveys both the network the packet is destined to and the host on the network. Now we have a general appreciation for the way source and destination IPv4 addresses are used to convey information, let's move up the protocol suite and examine how TCP and UDP are able to denote the application being transported. We will also examine the key differences between each transport layer protocol.

The transport layer

In the TCP/IP protocol suite, the transport layer is equivalent to layer 4 in the ISO Reference Model. While the Protocol field in the IPv4 header enables up to 256 higher layer protocols to be defined, the two transport layer protocols commonly used in the protocol suite are the TCP and the UDP.

Transmission Control Protocol (TCP)

The TCP represents a reliable, connection-oriented protocol. It is reliable because the protocol includes an error detection and correction capability. The protocol is connection oriented because it supports a three-way hand-shaking method under which the recipient must make its presence known prior to exchanging the data.

Figure 3.5 illustrates the format of the TCP header. From the viewpoint of content delivery, the source and destination ports are of key concern, since their values identify the application being transported.

Both source and destination port fields are 16-bits in length. Prior to discussing those ports, a few words are in order concerning the other fields in the TCP header. Let's quickly review a few of those fields to obtain an appreciation for why TCP is considered to represent a reliable, connection-oriented layer 4 protocol.

Sequence Number

The Sequence Number field is 32-bits in length. This field contains the sequence number of the first byte in the TCP segment unless the SYN bit

Source port		Destination port	
Sequence number			
Acknowledgement number			
Data	Reserved	Flags	Window
Checksum		Urgewnt pointer	
Options		Padding	

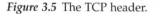

Figure 3.5 The TCP header.

(located in the flag field) is set. If the SYN bit is set, the sequence number becomes the initial sequence number (ISN) and the first data byte is ISN + 1.

Acknowledgment Number field

The Acknowledgment Number field is 32-bits in length. If the ACK control bit (located in the Flag field) is set, the Acknowledgment Number field contains the value of the next sequence number the sender of the TCP segment is expecting to receive. Once a connection is established, the next sequence number is always present in the Acknowledgment Number field. Thus, the Sequence Number and Acknowledgment Number fields not only provide a method for ensuring the correct order of segments at a receiver they also provide a way to note if a segment is lost.

Window field

The Window field is 16 bits in length. This field contains the number of data bytes beginning with the one indicated in the Acknowledgment Number field, which the sender of the segment is willing to accept. You can view the entry in the Window field as a flow control mechanism, since a small entry reduces the transmission of data per segment while a larger entry increases the amount of data transmitted per segment.

Checksum field

The Checksum field is similar to the Window field with respect to field length, since it is also 16 bits. The Checksum field contains the one's complement of the one's complement sum of all 16-bit words in the TCP header and text. If a TCP segment contains an odd number of bytes, an additional padded byte of zeros is added to form a 16-bit word for Checksum purposes; however, the padded byte is not transmitted as part of the TCP segment.

The Checksum also covers a 96-bit pseudo header that is conceptually prefixed to the TCP header. The pseudo header consists of the Source and Destination Address fields, the Protocol field, and the TCP length field. The purpose of the Checksum covering those fields is to provide protection against misrouted segments and enhances the reliability of this transport protocol. Now we have a general appreciation for TCP, let's turn to the second popular transport protocol in the TCP/IP protocol suite, UDP.

User Datagram Protocol (UDP)

The UDP represents a best effort, non-reliable transport protocol. Unlike TCP that requires the establishment of a connection prior to the transfer of data, when using UDP, data transfer occurs prior to knowing if a receiver is present. UDP relies on the application to determine if after a period of no response the session should terminate.

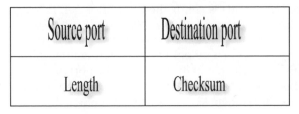

Figure 3.6 The UDP header.

Figure 3.6 illustrates the fields in the UDP header. Although both the TCP and UDP headers include 16-bit Source and Destination ports, the UDP header is streamlined in comparison to the TCP header. The UDP header has no flow control capability and, as we will shortly note, has a very limited error detection capability.

Length field

In examining the UDP header shown in Figure 3.6, the length field consists of 16-bits that indicate the length in bytes of the UDP datagram to include its header and data.

Checksum field

The Checksum is a 16-bit one's complement of the one's complement sum of a pseudo header of information from the prefixed IP header, the UDP header, and the data, padded with zeroed bytes, if necessary, to ensure a multiple of two bytes occurs. Similar to the TCP header Checksum provides protection against misrouted datagrams. However, unlike TCP there is no method within UDP for error detection of transmitted data, requiring the application to take charge of any required error detection and correction operation.

Port meanings

In our prior examination of TCP and UDP headers, we noted that layer 4 protocols have 16-bit Source and Destination fields. Because those fields function in the same way for each protocol, we will discuss their operation as an entity.

Destination Port

The Destination Port indicates the type of logical connection provided by the originator of the IP datagram. Here the term logical connection more specifically refers to the application or service transported by the TCP segment, which is identified by a port number in the Destination port field.

Source Port

The Source Port is normally set to a value of zero by the originator. However, when meaningful, the assignment of a non-zero value indicates the port of the sending process, which will then indicate the port to which a reply should be addressed. Because the values of each field are port numbers, this author would be remiss if he did not discuss their ranges.

Port numbers ranges

Each 16-bit Destination Port and Source Port field is capable of transporting a number from 0 through 65,535, for a total of 65,536 unique port numbers. Port numbers are divided into three ranges referred to as Well Known Ports, Registered Ports, and Dynamic or Private Ports. Well Known Ports are those port numbers from 0 through 1023, or the first 1024 port numbers. Registered Ports are those port numbers from 1024 through 49,151, while Dynamic or Private Ports are those port numbers from 49,152 through 65,535. Well Known Port numbers are assigned by the Internet Assigned Numbers Authority (IANA) and are used by system processes or by programs to identify applications or services. Table 3.1 lists some of the more common Well Known Port numbers. Although port numbers listed in Table 3.1 are applicable to both TCP and UDP, port numbers are commonly used with only one protocol. For example, FTP is transported as a reliable, connection-oriented process that occurs through TCP. In comparison, Simple Network Management Protocol (SNMP) is transported on a best-effort basis by UDP. However, some applications, such as Voice over IP use a combination of TCP and UDP. For example, when dialing a telephone number, the dialed digits are transported by TCP, which is a reliable protocol. However, once a connection is established to the dialed party, digitized voice is transported via UDP. The reason is real-time voice cannot be retransmitted if a bit error occurs. The application would either drop an errored packet or ignore the error when it reconstructs a short segment of voice transported via a UDP

Table 3.1 Common Well-Known Port Numbers

Port Number	Description
17	Quote of the Day
20	File Transfer Protocol-Data
21	File Transfer Protocol-Control
23	Telnet
25	Simple Mail Transfer Protocol
43	Whois
53	Domain Name Server

packet. Now we have an appreciation for the operation and utilization of the TCP/IP Transport Layer, we will conclude this chapter by turning to the Domain Name System (DNS). We will review how the DNS operates, not only to obtain an appreciation of how name resolution occurs but, to obtain the knowledge necessary to appreciate how DNS can be used as a way to support load balancing, a topic we will discuss in more detail later in this book.

3.2   The Domain Name System

When you enter a URL into your Web browser or send an email message, you more than likely use a domain name. For example, the URL http://www.popcorn.com contains the domain name popcorn.com. Similarly, the email address beverly@popcorn.com contains the same domain name.

Need for address resolution

While domain names are easy to remember, routers, gateways, and computers do not use them for addressing. Instead, computational devices are configured using dotted decimal digits to represent their IPv4 addresses.

Dotted decimal addresses are converted into binary and represent the true address of computational devices. Although many books reference IPv4 addresses as being assigned to computational devices, in actuality those addresses are assigned to device interfaces. This explains how routers and network servers with multiple network connections, can have packets transmitted from each interface with a distinct source address and receive packets with an explicit destination address that corresponds to a particular interface. Because IPv4 Class A, B, and C addresses indicate both a network and host address, such addresses identify both a network for routing purposes and a particular device on a network.

Using IPv4 addressing by computational devices means a translation device that resolves the domain name into an IP address is required to provide routers with information necessary to deliver packets to their intended destination. That translation or resolution service is referred to as the DNS and is the focus of this section.

Domain name servers

Computers used to translate domain names to IP addresses are referred to as domain name servers. There are a series of domain name servers that maintain databases of IP addresses and domain names, enabling a domain name to be resolved or translated into an IP address. Some companies operate a domain name server on their local area network while other organizations depend on the DNS operated by their ISP.

If a browser user enters a URL for which no previous IPv4 address was found, the local DNS on the organization's LAN will query the ISP's DNS

Table 3.2 Top-Level Domains

.aero	Aviation
.biz	Business organizations
.com	Commercial
.coop	Co-operative organizations
.edu	Educational
.gov	Government
.info	Information
.imt	International organization
.mil	U.S. Department of Defense
.museum	Museums
.name	Personal
.net	Networks
.org	Organizations

to determine if a resolution occurred at a higher level. Similarly, if the organization uses the services of the ISP's DNS, and a query does not result in a resolution, that DNS will forward the query to a higher authority. The highest authority is referred to as the top-level domain.

Top-level domain

Each domain name consists of a series of character strings separated by dots (.). The left-most string references the host, such as www or ftp. The right-most string in the domain name references the top-level domain such as gov or com.

When the Internet was initially established, there were only a handful of top-level domains. Those top-level domains included .com (commercial), .edu (educational), .gov (government), .mil (U.S. Department of Defense), .net (networks), and .org (organization). Since then, domain name registries were considerably expanded, as were top-level domain name servers. Table 3.2 lists presently defined domain name registries other than those defined for countries. Concerning the registries defined by country, there are presently over 100 two-letter domain registries for countries, such as .ar (Argentina), .il (Israel), and .uk (United Kingdom). The IANA is responsible for defining domain name suffixes.

Within each top-level domain, there can be literally tens of thousands to millions of second-level domains. For example, in the .com first level domain you have:

- Microsoft
- Google
- Yahoo!

and millions of other entries. Although every .com top-level domain must be unique, there can be duplication across domains. For example, lexus.com and lexus.biz represent two different domains. By prefixing a word to the domain, such as ftp.popcorn.com or www.popcorn.com you obtain the name of a specific host computer in a domain. That computer has an IP address, which is determined through the DNS.

DNS operation

When you enter a URL into your browser in the form of a domain name, that name must be converted into an IP address. The browser will use that address to request a Web page from the computer whose interface is assigned that address. To obtain that address, the browser must use the facilities of a domain name server. Thus, the browser must know where to look to access the name server.

Configuring your computer

When you install your computer's TCP/IP software, one of the first functions you need to perform is to configure your network settings. You will set your computer's IP address, its subnet mask, the default gateway, and the address of the name server your computer should use when it needs to convert domain names to IP addresses.

Figure 3.7 illustrates the Microsoft Windows 2000 Internet Protocol (TCP/IP) Properties dialog box. Note if you select the button Use the following IP addresses, you are able to specify the IP address, subnet mask, default gateway, and up to two DNS server addresses. However, if your organization uses the Dynamic Host Configuration Protocol (DHCP), you would select the button labeled Obtain an IP address automatically, which would result in the DNS addresses being transmitted to the host from a DHCP server along with its IP address, subnet mask, and gateway address when the host connects to the network.

If you're working in a Windows environment there are several tools you can consider using to obtain DNS and other addressing information. If you're using an older version of Windows, such as WIN95 or WIN98, you can view current IP address assignments through WINIPFG.EXE. If you're using Windows 2000 or Windows XP, you can use ipconfig from the command prompt.

The top portion of Figure 3.8 illustrates using the ipconfg program without any options. When used in this way, the program returns the connection specific DNS suffix, IP address, subnet mask, and default gateway address. Next, the ipconfig program was executed a second time, however,

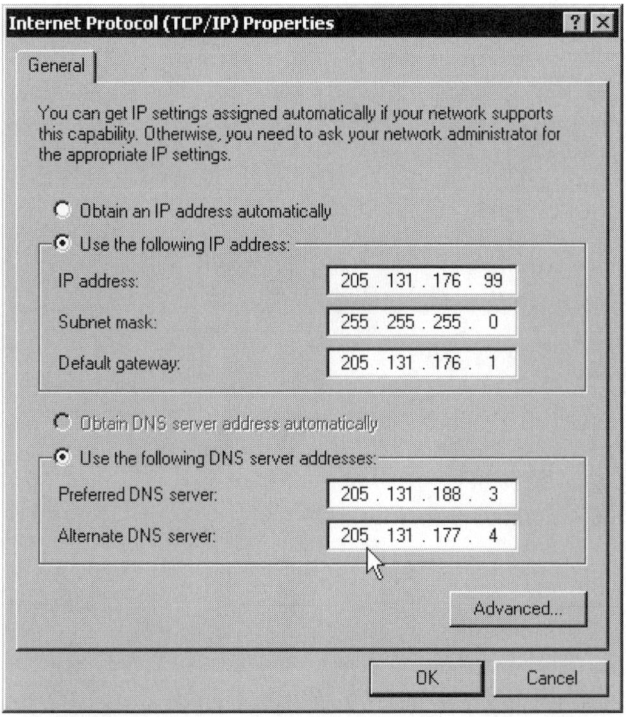

Figure 3.7 Using the Windows 2000 Internet Protocol (TCP/IP) properties dialog box.

this time the all option was included in the command line. Note using the all option provides additional information about the configuration of the computer to include the DHCP and DNS server addresses and DHCP leasing information.

Once a computer knows the IP address of its domain name server, it can request the server to convert a domain name into an IP address. If the name server received a prior request to obtain an IP address for a host with a particular domain name, such as www.popcorn.com, the server merely needs to access its memory to return the IP address associated with the request previously stored in cache memory to the computer making the resolution request. If the name server did not have prior knowledge of the resolution, it would then initiate contact with one of the root name servers.

Root name servers

Currently there are 13 root name servers in existence, with most of them located in the United States while several servers are located in Japan and London. Each root server functions in a similar way, responding to a DNS query with the address of a name server for the top-level domain for a particular query. That is, each root server knows the IP address for all of the name servers that support a particular top-level domain. Thus, if your

```
Command Prompt                                                          _|d| X|
Microsoft Windows 2000 [Version 5.00.2195]
(C) Copyright 1985-2000 Microsoft Corp.

h:\>ipconfig

Windows 2000 IP Configuration

Ethernet adapter Local Area Connection:

        Connection-specific DNS Suffix  . : opm.gov
        IP Address. . . . . . . . . . . . : 172.26.22.213
        Subnet Mask . . . . . . . . . . . : 255.255.255.0
        Default Gateway . . . . . . . . . : 172.26.22.1

h:\>ipconfig/all

Windows 2000 IP Configuration

        Host Name . . . . . . . . . . . . : MCN_3393708
        Primary DNS Suffix  . . . . . . . : opm.gov
        Node Type . . . . . . . . . . . . : Hybrid
        IP Routing Enabled. . . . . . . . : No
        WINS Proxy Enabled. . . . . . . . : No
        DNS Suffix Search List. . . . . . : opm.gov

Ethernet adapter Local Area Connection:

        Connection-specific DNS Suffix  . : opm.gov
        Description . . . . . . . . . . . : Intel(R) PRO/100 VE Network Connecti
        Physical Address. . . . . . . . . : 00-07-E9-ED-09-0F
        DHCP Enabled. . . . . . . . . . . : Yes
        Autoconfiguration Enabled . . . . : Yes
        IP Address. . . . . . . . . . . . : 172.26.22.213
        Subnet Mask . . . . . . . . . . . : 255.255.255.0
        Default Gateway . . . . . . . . . : 172.26.22.1
        DHCP Server . . . . . . . . . . . : 172.26.32.7
        DNS Servers . . . . . . . . . . . : 172.26.32.7
                                            172.20.8.8
        Primary WINS Server . . . . . . . : 172.26.32.7
        Secondary WINS Server . . . . . . : 172.20.8.8
        Lease Obtained. . . . . . . . . . : Thursday, October 28, 2004 12:51:28
        Lease Expires . . . . . . . . . . : Thursday, November 04, 2004 11:51:28

h:\>
```

Figure 3.8 Using ipconfig to obtain information about the network settings associated with a computer.

browser was pointed to the URL www.popcorn.com, and the local name server had not previously resolved the IP address for that host and domain address, the local name server would contact one of the root name servers. Assuming the root name server had not previously resolved the host and domain name, the root server would respond with the IP address of the domain, which in our example is the name server for the .com domain, enabling your name server to access that server.

Root servers are labeled A through M, with each name server having a file that contains information in the form of special records that contain the

name and IP address of each root server. Each root server in turn is configured with the IP addresses of the name servers responsible for supporting
the various top-level domains. Thus, a resolution request that cannot be
serviced by the local domain server is passed to an applicable root server,
which in turn returns the IP address of the top-level domain to the local
DNS. That name server then transmits a query to the top-level name server,
such as a COM, EDU or GOV name server, requesting the IP address for the
name server for the domain in which the host that requires address resolution
resides. Because the top-level domain name server has entries for all domain
servers for its domain, it responds with the IP address of the name server
that handles the domain in question. The local name server uses that IP
address to directly contact the name server for the host and domain name
it needs to resolve, such as www.popcorn.com. That name server returns the
IP address to the local name server, which then returns it to the browser. The
browser then uses that IP address to contact the server for www.popcorn.com
to retrieve a Web page.

The NSLOOKUP tool

If you're using a version of Microsoft Windows that has the TCP/IP protocol
suite installed, you can use the nslookup diagnostic tool to obtain information from domain name servers. Nslookup operates in two modes: interactive
and noninteractive. The noninteractive mode is normally used when you
need to obtain a single piece of data, while the interactive or program mode
lets you issue a series of queries.

The top portion of Figure 3.9 illustrates using nslookup in its noninteractive mode. In this example, nslookup was used to obtain the IP address
of the Web server www.eds.com. In the second example, nslookup was
entered by itself to place the program into its interactive mode of operation.
The program responds by indicating the name server and its IP address and
then displays the > character as a prompt for user input. You can now enter
an nslookup command. In the interactive example, the set all command was
entered to obtain a list of set options. Note that one option is the root server,
with its current value set to A.ROOT-SERVERS.NET. This represents the root
server the local DNS server uses by default. Next, the server command was
used to set the root server as the default name server. Note this action
returned the IP address of the root server.

Expediting the name resolution process

Name servers use caching to expedite the name resolution process. When
a name server resolves a request, it caches the IP address associated with
the name resolution process. The next time the name server receives a request
for a previously resolved domain, it knows the IP address for the name
server handling the domain. The name server does not have to query the
root server since it previously learned the required information and can

```
Command Prompt - nslookup                                          _ □ X
C:\>nslookup www.eds.com
Server:  ns3.opm.gov
Address:  205.131.188.3

Non-authoritative answer:
Name:     eagle-ldir.xweb.eds.net
Address:  207.37.253.223
Aliases:  www.eds.com


C:\>nslookup
Default Server:  ns3.opm.gov
Address:  205.131.188.3

> set all
Default Server:  ns3.opm.gov
Address:  205.131.188.3

Set options:
  nodebug
  defname
  search
  recurse
  nod2
  novc
  noignoretc
  port=53
  type=A
  class=IN
  timeout=2
  retry=1
  root=A.ROOT-SERVERS.NET.
  domain=
  MSxfr
  IXFRversion=1
  srchlist=

> server a.root-servers.net
Default Server:  a.root-servers.net
Address:  198.41.0.4

>
```

Figure 3.9 Using Microsoft's nslookup in noninteractive and interactive query mode.

simply retrieve it from cache memory. While caching can be very effective, it has two limitations. First, not all requests are duplicates of prior requests. Secondly, cache memory is finite. When a name server receives an IP address, it also receives a TTL value associated with the address. The name server will cache the address until the TTL period expires, after which the address is purged to make room for new entries.

DNS resource records

The key to the operation of name servers are resource records. Resource records define data types in the DNS. DNS records are coded in American Standard Code for Information Interchange (ASCII) and translated into a binary representation for internal use by a DNS application.

The DNS system defines a number of Resource Records (RRs). The text representation of RRs are stored in what are referred to as zone files that can be considered to represent the domain name database.

Start of Authority (SOA) resource record

At the top of each zone file is a Start of Authority (SOA) record. This record identifies the zone name, an email contact, and various time and refresh values

Table 3.3 The SOA Record Format and an Example of Its Use

```
SOA Format
DOMAIN.NAME.   IN   SOA   Hostname.Domain.Name. Mailbox.Domain.Name. (
                          1        ; serno (serial number)
                          86400    ; refresh in seconds (24 hours)
                          7200     ; retry in seconds (2 hours)
                          259200 ; expire in seconds (30 days)
                          345600 ; TL in seconds (4 days)
SOA Record Example
POPCORN.COM    IN   SOA   POPCORN.COM. gheld.popcorn.com. (
                          24601    ; serial number
                          28800    ; refresh in 8 hours
                          7200     ; retry in 2 hours
                          259200 ; expire in 30 days
                          86400    ; TTL is 1 day
```

applicable to the zone. The top portion of Table 3.3 illustrates the RFC 1537 defined format for the SOA record while the lower portion of that illustration shows an example of the SOA record for the fictional popcorn.com domain.

In examining the format of the SOA record shown in Table 3.3, the trailing dot (.) after the domain name signifies that no suffix is to be appended to the name. The class of the DNS record is shown as IN, which stands for Internet, while SOA indicates the type of DNS record. The mailbox is of the individual responsible for maintaining DNS for the domain. The serial number (serno) indicates the current version of the DNS database for the domain and provides the way other name servers can note that the database was updated. The serial number commences at one and is increased by one each time the database changes.

The Refresh entry tells the secondary name server how often to poll the primary name server for changes. The Retry entry defines the interval in seconds at which the secondary name server tries to reconnect to the primary name server in the event it failed to connect at the Refresh interval.

The Expire entry defines how long the secondary name server should use its current entry if it is unable to perform a Refresh while the TTL value applies to all records in the DNS database on a name server.

Name Server (NS) Records

While there is only one SOA record per domain, because there can be multiple name servers there can be multiple NS records. Name servers use NS records to locate one another and there must be at least two NS records in every DNS entry. The format of an NS record is shown below:

```
DOMAIN.NAME. IN NS Hostname.Domain.Name..
```

Address (A) records

The purpose of the Address (A) record is to map the host name of a computer to its numeric IP address. The format of an Address record is indicated below:

```
Host.domain.name.  IN  A  www.xxx.yyy.zzz
```

Host Information (HINFO) record

A Host Information (HINFO) record is optional. When used, it can provide hardware operating system information about each host. The format of an HINFO record is shown below:

```
Host.DOMAIN.NAME.  IN  HINFO  "cputype"  "OS"
```

Mail Exchange (MX) records

The purpose of a Mail Exchange (MX) record is to allow mail for a domain to be routed to a specific host. A host name can have one or more MX records since large domains will have backup mail servers. The format of an MX record is shown below:

```
Host.domain.name. IN MX MM otherhost.domain.name.
                  IN MX MM otherhost2.domain.name.
```

The preference numbers (NN) signify the order in which mailers will select MX records when attempting mail delivery to the host. The lower the number the higher the host is in priority.

To illustrate the MX record, assume your organization's mail server had the address mail.popcorn.com. Further, assume you wanted your email addresses to be user@popcorn.com. rather than user@mail.popcorn.com. To accomplish this, your MX record would be coded as follows:

```
popcorn.com.  IN MX 10 mail.popcorn.com.
```

Carnonical Name (CNAME) Records

The Carnonical Name (CNAME) record enables a computer to be referred to by an alias host name. The CNAME record format is shown below:

```
Alias.domain.name.  IN CNAME otherhost.domain.name.
```

It's important to note that there must be an A record for the host prior to adding an alias. The host name in the A record is known as the carnonical or the official name of the host.

Other records

In addition to the previously mentioned DNS records, there are many other resource records. Those records range in scope from Pointer (PTR) Records that provide an exact inverse of an A record allowing a host to be recognized by its IP address to an A6 resource record, which is used for an IPv6 address for a host. A full list of DNS record types can be obtained from IANA DNS parameter listings.

chapter four

The CDN Model

Until now, we briefly covered the major reasons for having a content delivery system and a few of the methods that could facilitate the operation of a content delivery network (CDN). In this chapter, we will probe deeper into the content delivery model, examining the so-called edge operations that move the content of customers to the edges of the Internet. In actuality, the term edge while appropriate here, can mean different things to different persons with respect to their physical location. However, prior to discussing edge operations we will first build on information previously presented in this book to obtain a better appreciation for the rationale for CDN. We will view the Internet as a transmission facility that has a series of critical links to understand bottlenecks and how a distributed CDN can overcome such bottlenecks. Because this author believes that both sides of a coin need to be shown, as we discuss edge operations we will also examine some of the limitations associated with the distribution of content across the Internet.

4.1 Why Performance Matters

Earlier in this book, we looked at the interconnection of communication carrier networks at peering points, and how those locations could adversely affect the flow of data. In addition, we looked at the flow of traffic from users distributed across the globe accessing a common Web server and noted that some users had their traffic flow over a large number of router hops to reach the server. Because router hops and the crossing of traffic at peering points correspond to transmission delays, as the distance between the user of the server and the server increases so too will the delays. Those delays can affect the ability of potential customers of Web sites, and the resulting economics associated with poor performance is worth noting.

Economics of poor performance

To understand the economics associated with poor performance, let's assume your organization's Web site sells discount airline seats. Now let's assume a potential customer of the Web site enters his or her dates of travel and travel locations to request a discount price. However, in doing so the potential customer may then wish to alter their travel dates and check the rates to one or more alternate but nearby locations since airlines are notorious for having many travel restrictions as well as having rates between cities that enable considerable savings by flying either from or to a closely located airport from the intended point of origination or destination. Because of the travel discrepancies, the typical potential customer may need to enter data on a series of Web pages and flip through a series of page responses.

If the potential customer is comparing prices of several travel sites, they are typically pressed for time. A potential customer who experiences page access and display delays associated with router hop and peering points may in effect bail out from further accessing the Web site for which they are experiencing delays and proceed to another site.

In the wonderful world of marketing, we are probably familiar with the adage time is money. We can equate this adage to the Internet by noting that every bail out of a potential customer results in the potential loss of revenue. However, when a customer is driven away because of what they perceive as a poor performing Web site, but is in reality the delays resulting from the path through which data flows between the potential customer and the Web site, a lasting impression of poor site performance can occur. This means the Web site operator not only loses the possibility of a sale, he or she may lose the opportunity for future sales because the potential customer now has a negative view of the Web site.

Predictability

Predictability is another problem potential customers encounter when accessing a Web server across many router hops and peering points. A centrally hosted infrastructure results in potential customers at locations from Andorra to Zambia accessing a common location. Some routes may require the traversal of a large number of router hops and peering points, while other routes may require the traversal of a fewer number of router hops and perhaps a few or even no peering points. As you can imagine, depending on the geographic location of the potential customer and Web server being accessed, different potential customers can be expected to encounter different delays, making the access and retrieval of Web pages anything but predictable. In addition, the geographic distribution of potential customers also needs consideration regarding the time of day when Web access occurs.

Dataflow on the Internet has peaks and valleys that partially correspond to the typical workday. For Monday through Friday activity increases from a relatively low level prior to the 8 A.M. to noon period as workers arrive and perform both job related and personal activities that require Internet access. From approximately noon until 2 P.M., activity tapers off as workers go to lunch and run errands. In the afternoon, activity peaks between 4 P.M. and 5 P.M. and then tapers off as people leave work. However, as workers arrive at home, some persons begin to use the Internet to perform a variety of activities that may not be possible or are restricted at work, ranging from checking personal email to retrieving stock market quotations. There are several activity peaks during the late afternoon and evening. In addition, on weekends when there is minimal access of the Internet from work, tens of millions of persons located around the globe access the Internet from home or from libraries, colleges, and universities creating variable traffic peaks and valleys throughout the weekend.

When traffic loads are considered along with the geographic location of potential customers, it becomes apparent that the potential customer of a centrally located computer infrastructure consisting of a Web server and backend database servers will have different experiences with Web server access and page retrieval operations each time they point their browser to a particular Web site. This unpredictability can result in the loss of potential customers who decide to perform their search for products on other Web sites, resulting in an additional effect on the bottom line of the Web site operator.

Customer loyalty

During the initial Internet boom from 1997 through the year 2000, many market research organizations viewed the popularity of sites only with respect to page clicks. In the euphoria of that period, the fact that only a few clicks were converted into purchases was irrelevant. Fortunately, the burst of the so-called Internet bubble resulted in a return to rational market research where the bottom line matters.

When a potential customer cannot predictably access an organization's Web site, they normally make a rational decision to go elsewhere. This decision can occur via the entry of a new URL to using a search engine to locate another site providing a similar product. Over time, the potential customer may even remove the current site from their browser's Favorites list. Regardless of the action performed, the result will be similar because instead of acquiring a loyal customer who could be the source of repeated business the potential customer is driven away.

Scalability

Another problem associated with the centralized Web site model is scalability. The centralized model requires the Web site operator to add more equipment to one location to satisfy access increases that can occur from locations scattered over the globe.

In a distributed model where content delivery is moved to the literal edges of the Internet, an increase in Web access is distributed over many servers. As a result, it may not be necessary to upgrade any Web server. In addition, if the organization is using the facilities of a CDN provider, the responsibilities associated with computer upgrades become the responsibility of the service provider. This means if your organization enters into a contract with a service provider that includes a detailed service level agreement, the service provider will upgrade their equipment at one or more locations when service becomes an issue. This upgrade at most should only temporarily affect one of many locations and should be compared to a central site upgrade that can affect all customers on a global basis. The scalability issue also includes the effect on potential customers, with a centralized model providing a high degree of a complete outage occurring during an equipment upgrade process while a distributed model upgrade only affects potential customers whose traffic flows through a particular edge server.

Flexibility

The centralized approach can result in a complete outage during a hardware upgrade. A distributed approach where content is moved forward to edge servers provides more flexibility with respect to hardware upgrades. First, moving content onto many servers can forgo the need for a centralized hardware upgrade. Secondly, if a hardware upgrade becomes necessary, at most it only affects one of many edge servers at a time and allows your organization to better plan for partial outages. Using a CDN can provide a more flexible solution to your Web site data access requirements.

Company perception

There is the well known adage you are what you eat. In the wonderful world of the Internet, the performance of your company's Web site can have a significant influence on how actual and potential customers view your organization and your organization's brand perception. If users attempting to access your corporate Web site encounter significant delays, not only will this result in a number of users pointing their browsers elsewhere, it will result in a negative view of your organization. Unfortunately, a heavy investment in commercial advertising can come to naught if customers and potential customers decide to abandon your organization's Web site. The performance of your organization's Web site including access delays can have a direct impact on brand perception and customer loyalty.

Summary

Based on information presented in this section, it's obvious that Web performance matters. Because customers and potential customers consider access delays to represent Web server performance, the centralized server model has many limitations. Those limitations include a lack of predictability, the potential loss of customer loyalty, difficulty in scaling a centralized server to accommodate traffic growth, an impairment of organizational flexibility, and the possibility that the operation of a centralized site will provide delays that negatively impact the perception of your organization. Because these limitations can result in the loss of customer revenue, there is an economic penalty associated with the limitations. That economic penalty can vary considerably, based on the severity of the limitations and the type of merchandise or service sold by the Web site.

Prior to examining how moving content to edge servers can reduce latency and enhance server access, let's obtain a more detailed view of the factors associated with Web server access delays. Let's turn to examining Internet bottlenecks to obtain a better understanding behind the rationale for moving content toward actual and potential users.

4.2   Examining Internet Bottlenecks

Previously in this book, we noted that the distance between the user and a Web site in terms of the number of router hops and peering points significantly contributes to site access delays. In this section, we will probe deeper into Internet bottlenecks and examine data flow from the source to the destination, noting the effect of a series of potential and actual bottlenecks on Web server access.

Entry and egress considerations

Two of the often overlooked Internet bottlenecks are the entry and egress transport facilities used to access a particular Web server.

The entry transport facility refers to the type of access the user employs to access the Internet and the activity over that transport facility. While the egress transport facility refers to the communications link from the Internet to a particular Web server, it also represents a reverse connection. The egress transport facility with respect to a user becomes the access transport facility of the server. Similarly, the entry transport facility of the user can be viewed as the egress transport facility of the server. To eliminate possible confusion, we can note that the typical browser user sends requests in the form of URLs to a Web site to retrieve a Web page. The Web server responds with a Web page whose size in bytes is normally several orders of magnitude greater than the packet sent by the browser user. Instead of focusing on the transmission of the URL to the server, we can focus on the Web server's response. We can note the user's access method to the Internet results in a delay in the delivery of a server page based on the operating rate of the access line. Similarly, the user's egress transport facility can be viewed as a delay mechanism with respect to the server delivering a Web page to the Internet. Now we have an appreciation for the way data flows from the source to the destination, we can focus on Web page delays with respect to the browser user's view of access and egress.

Access delays

The access line connecting a browser user to the Internet represents the egress delay associated with delivering a Web page to the user. Because there are several types of transport facilities a browser user can use to access the Internet, we need to consider each method when computing the effect of the access transport facility on Web page egress delays. For example, a user might access the Internet via dial-up connection using the public switched telephone network at 56 Kbps, over a Digital Subscriber Line (DSL) modem connection at 1.5 Mbps, a cable modem connection operating at 6 Mbps, or a corporate T1 connection operating at 1.544 Mbps. While the first three connection methods provide dedicated access to the Internet, the corporate T1 connection represents a shared access method, with achievable throughput based on the number of corporate users accessing the Internet and their activity. Assuming each corporate user is performing a similar activity and 10 users are accessing the Internet, the average throughput of each user becomes 1.544 Mbps/10 or 154,400 bps.

In actuality, a T1 line that operates at 1.544 Mbps uses 8000 bits per second for framing. The actual data rate available to transport data over a T1 connection becomes 1.544 Mbps — 8000 bps or 1.536 Mbps. For most organizations estimating T1 performance, the variability of an estimate within a margin of error of 10 to 20 percent allows computations to occur using a data rate of 1.544 Mbps. However, as noted, a better measurement occurs when the data transport capacity of 1.536 Mbps is used for a T1 line.

Most Internet entry actions consist of transmitting a short URL to access a server page. Throughput delays associated with requesting a Web page do not significantly vary among these access methods. However, the opposite is not true. There can be significant differences in Web page display delays based on the method a user employs to access the Internet. For example, consider Table 4.1 where the delay or latency associated with delivering a Web page vary in size from 10000 bytes to 300,000 bytes in increments of 10,000 bytes based on four data rates.

In examining the entries in Table 4.1, let's start with the leftmost column, which is the Web page size column. Most Web pages contain a mixture of text and graphics, with the graphics primarily in the Joint Photographic Experts Group (JPEG) format that permits a high degree of image compression. Even so, it's common for a typical Web page to consist of between 150,000 and 175,000 bytes. One notable exception to this average Web page size is the Google home page shown in Figure 4.1. Note the Google home page is streamlined, with only one graphic image on the page. This facilitates the delivery of that home page to users regardless of the data transport mechanism they are using to access the Internet.

At the opposite end of Web page design with respect to graphic images are the home pages of the major television networks, such as abc.com, cbs.com, fox.com, and nbc.com, and portals, such as Yahoo.com and

Table 4.1 Web Page Delays (in Seconds) Based on Page Size in Bytes and the Access Line Connection

Web Page Size	56000 bps	150000 bps	1.544 Mbps	6 Mbps
Bytes	Data rate delay	Data rate delay	Data rate delay	Data rate delay
10000	1.42857	0.53333	0.05208	0.01333
20000	2.85714	1.06667	0.10417	0.02667
30000	4.28571	1.60000	0.15625	0.04000
40000	5.71429	2.13333	0.20833	0.05333
50000	7.14286	2.66667	0.26042	0.06667
60000	8.57143	3.20000	0.31250	0.08000
70000	10.00000	3.73333	0.36458	0.09333
80000	11.42857	4.26667	0.41667	0.10667
90000	12.85714	4.80000	0.46875	0.12000
100000	14.28571	5.33333	0.52083	0.13333
110000	15.71429	5.86667	0.57292	0.14667
120000	17.14286	6.40000	0.62500	0.16000
130000	18.57143	6.93333	0.67708	0.17333
140000	20.00000	7.46667	0.72917	0.18667
150000	21.42857	8.00000	0.78125	0.20000
160000	22.85714	8.53333	0.83333	0.21333
170000	24.28571	9.06667	0.88542	0.22667
180000	25.71429	9.60000	0.93750	0.24000
190000	27.14286	10.13333	0.98958	0.25333
200000	28.57143	10.66667	1.04167	0.26667
210000	30.00000	11.20000	1.09375	0.28000
220000	31.42857	11.73333	1.14583	0.29333
230000	32.85714	12.26667	1.19792	0.30667
240000	34.28571	12.80000	1.25000	0.32000
250000	35.71429	13.33333	1.30208	0.33333
260000	37.14286	13.86667	1.35417	0.34667
270000	38.57143	14.40000	1.40625	0.36000
280000	40.00000	14.93333	1.45833	0.37333

Table 4.1 (continued) Web Page Delays (in Seconds) Based on Page Size in Bytes and the Access Line Connection

Web Page Size	56000 bps	150000 bps	1.544 Mbps	6 Mbps
bytes	data rate delay	data rate delay	data rate delay	data rate delay
290000	41.42857	15.46667	1.51042	0.38667
300000	42.85714	16.00000	1.56250	0.40000

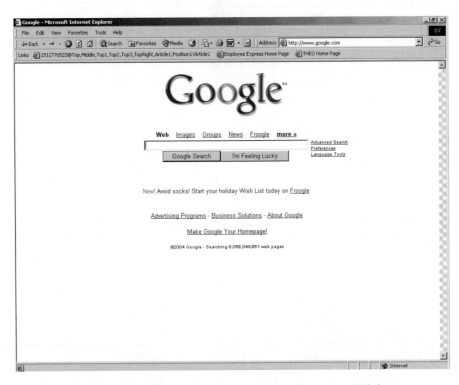

Figure 4.1 The Google home page is optimized for delivery to a Web browser user.

msnbc.com. In addition, many newspapers' online Web sites, such as nytimes.com and jpost.com are packed with a large number of small graphic images that cumulatively results in a Web page that can easily exceed 150,000 bytes of data. For example, consider Figure 4.2, which shows the home page of the *New York Times*. If you scroll down the remainder of the home page of the *New York Times*, you will encounter a series of small graphic images that cumulatively add up to produce a rather large download.

You can obtain an appreciation for the size of graphic images by moving your cursor over an image and performing a right click operation. From the resulting pop-up menu, select Properties — this will display the size of the image in bytes and other information about the image.

Figure 4.2 The *New York Times* on the Web home page contains graphics scattered on the page that cumulatively result in a large download of bytes of data.

Figure 4.3 illustrates an example of the display of the Properties box associated with the image located to the right of the box. Note this small image, which is one of approximately ten to a dozen small images on the typical home page of the *New York Times* for any day of the week, consists of almost 13,000 bytes. When you add up the number of bytes for the series of small images and other graphics on the typical home page of the *New York Times*, the amount of data easily approaches 150,000 bytes.

Returning to Table 4.1, let's focus on column 2 through column 5. Those columns indicate the delay in seconds for the reception of a Web page of indicated size in bytes based on the transport facility employed by the browser user. Focus on the columns associated with the Web page size of 150,000 bytes that represents an average Web page size. We can note that the page exit delay or latency will range from approximately 21 seconds when the user employs a 56 Kbps dial-up modem to a fifth of a second when a cable modem operating at 6 Mbps is used. Similarly, at a Web page size of 200,000 bytes the Web page delays range from approximately 28.6 seconds at 56 Kbps to 0.266 seconds at a cable modem operating rate of 6 Mbps. As you might expect, Web page delays associated with exiting the Internet via the browser user's access line increase as the Web page size increases. In addition, the delay is proportional to the operating rate of the access line.

Figure 4.3 Viewing the Properties box associated with a small image on the *New York Times* home page.

Both the page size in bytes returned to the user and his or her Internet connection method plays an important role in the overall delay.

Now that we have an appreciation for the delays attributable to the type of access to the Internet employed by browser users, let's turn to the egress delays. We are referring to the connection from the Internet to a particular Web server. Because a Web page is several orders of magnitude larger in terms of bytes than a URL request, we can ignore the delay associated with the packet containing the URL flowing to the Web server. Instead, we can turn to the delay that occurs as the requested page flows over the egress transport facility employed by the browser user to access the Web server. We can continue to focus on the flow of Web pages towards the user.

Egress delays

Similar to the way a browser user's connection to the Internet has a bearing on latency, so does the connection of the Web site to the Internet. Because input delays are based on the size of the data packet containing a small amount of data in the form of a URL request, we can ignore input delays. Similar to our investigation of user access delays, the primary egress transport facility delay involves the flow of a Web page over the transport facility that connects a Web site to the Internet.

Most Web sites are connected to the Internet by T1 or T3 lines, with the T3 lines operating at approximately 45 Mbps. However, unlike our prior computations for the Internet access line that is normally not shared, the egress connection is shared by many users. We need to consider the average number of users accessing a Web server, as each user request results in the return of a Web page. While it's true that different users will be requesting different Web pages formed from different text and graphic images, for simplicity we can consider each Web page has a similar composition in bytes. We can modify our previous computations performed in Table 4.1 to reflect the activity of additional users at a Web site.

Table 4.2 provides a summary of an Excel spreadsheet model to project the time delays in seconds associated with a Web page delivery. Similar to Table 4.1, the first column indicates varying Web page sizes in bytes, ranging from 10,000 bytes to 300,000 bytes in increments of 10,000 bytes. Columns 2, 3, and 4 indicate the delays associated with one, ten, and 100 users querying a server and sharing a 1.5 Mbps T1 transport facility connecting the server to the Internet. Similarly, columns 5, 6, and 7 indicate the delays associated with one, ten and 100 users accessing a common Web server and sharing a T3 transport facility operating at approximately 45 Mbps that connects the server to the Internet.

In examining the entries in Table 4.2, note that the delay in delivering a Web page from a server onto the Internet is primarily a function of three factors. Those factors include the Web page size in bytes, the transport facility operating rate, and the number of users requesting the delivery of Web pages over the common connection between the server and the Internet.

A careful examination of the data presented in Table 4.2 indicates that a popular Web site that has a T1 connection to the Internet can encounter significant Web page delivery delays when even 10 users are actively requesting Web pages. For example, at a Web page size of 150,000 bytes the delay in placing one Web page onto the Internet is eight seconds. When the number of users requesting Web pages increase to 100, the delay increases to 80 seconds, which is obviously a significant amount of time.

To reduce delay times many organizations upgraded their facilities by installing T3 connections to the Internet that operate at approximately 45 Mbps. If you examine the row in Table 4.2 associated with a Web page the size of 150,000 bytes and move to the rightmost column, you will note that using a T3 transmission facility when there are 100 users results in a Web page delay of approximately 2.67 seconds. While this is significantly less than the 80 seconds associated with the use of a T1 line shared by 100 users, it still represents a large delay if a user needs to scroll through a series of Web pages. In addition, we need to note that the access and egress transport facilities are cumulative, further adding to the delays experienced by a browser user accessing a Web server.

Table 4.2 Time Delays in Seconds for Delivering a Web Page from a Server to the Internet

Web Page Size	150000bps	150000bps	150000bps	45mbps	45mbps	45mbps
	Data rate	Data rate	Data rate	Data rate	Data rate	Data rate
bytes	1 user delay	10 user delay	100 user delay	1 user delay	10 user delay	100 user delay
10000	0.05333	0.53333	5.33333	0.00178	0.01778	0.17778
20000	0.10667	1.06667	10.66667	0.00356	0.03556	0.35556
30000	0.16000	1.60000	16.00000	0.00533	0.05333	0.53333
40000	0.21333	2.13333	21.33333	0.00711	0.07111	0.71111
50000	0.26667	2.66667	26.66667	0.00889	0.08889	0.88889
60000	0.32000	3.20000	32.00000	0.01067	0.10667	1.06667
70000	0.37333	3.73333	37.33333	0.01244	0.12444	1.24444
80000	0.42667	4.26667	42.66667	0.01422	0.14222	1.42222
90000	0.48000	4.80000	48.00000	0.01600	0.16000	1.60000
100000	0.53333	5.33333	53.33333	0.01778	0.17778	1.77778
110000	0.58667	5.86667	58.66667	0.01956	0.19556	1.95556
120000	0.64000	6.40000	64.00000	0.02133	0.21333	2.13333
130000	0.69333	6.93333	69.33333	0.02311	0.23111	2.31111

140000	0.74667	7.46667	74.66667	0.02489	0.24889	2.48889
150000	0.80000	8.00000	80.00000	0.02667	0.26667	2.66667
160000	0.85333	8.53333	85.33333	0.02844	0.28444	2.84444
170000	0.90667	9.06667	90.66667	0.03022	0.30222	3.02222
180000	0.96000	9.60000	96.00000	0.03200	0.32000	3.20000
190000	1.01333	10.13333	101.33333	0.03378	0.33778	3.37778
200000	1.06667	10.66667	106.66667	0.03556	0.35556	3.55556
210000	1.12000	11.20000	112.00000	0.03733	0.37333	3.73333
220000	1.17333	11.73333	117.33333	0.03911	0.39111	3.91111
230000	1.22667	12.26667	122.66667	0.04089	0.40889	4.08889
240000	1.28000	12.80000	128.00000	0.04267	0.42667	4.26667
250000	1.33333	13.33333	133.33333	0.04444	0.44444	4.44444
260000	1.38667	13.86667	138.66667	0.04622	0.46222	4.62222
270000	1.44000	14.40000	144.00000	0.04800	0.48000	4.80000
280000	1.49333	14.93333	149.33333	0.04978	0.49778	4.97778
290000	1.54667	15.46667	154.66667	0.05156	0.51556	5.15556
300000	1.60000	16.00000	160.00000	0.05333	0.53333	5.33333

Benefits of edge servers

The distribution of Web server content onto edge servers can significantly reduce many of the delays computed in Table 4.2. Because a lower number of browser users can be expected to access distributed Web sites in comparison to the number of users who can be expected to access a centralized site. Unfortunately, the distribution of server content onto edge servers will have no bearing on the delays associated with a user's access line. There are certain limitations associated with the distribution of server content, which will not significantly improve operations over the use of a centralized Web site. Now that we have an appreciation for Internet entry and egress delays due to transport facility operating rate and potential user sharing the facility. Let's turn to several additional bottlenecks. One bottleneck we previously covered that deserves a more detailed examination involves peering points and their effect on Web servers responding to a browser user's request.

Peering points

The Internet represents a collection of networks that are interconnected to permit the flow of data from a computer located on one network to a computer connected to another network. The networks that are interconnected, can range in size from a LAN with a handful of attached computers to the LAN, to networks operated by an Internet Service Provider (ISP) that can consist of hundreds of thousands of Digital Subscriber Line (DSL) or cable modem users or even large ISPs, such as America Online, which supports approximately 20 million users.

Rationale

Because of the global nature of the Internet, most networks are not directly connected to one another. Instead, networks were originally interconnected via the transmission facilities of one or more third party networks. An example of this interconnection is illustrated in Figure 4.4. In this example, for data transmitted from a computer user located on network A to arrive at a computer located on network E, the transmission facilities of two other networks, such as networks B and C or networks B and D must be employed to provide an interconnection to the destination network. Similarly, a computer user on network B who requires access to a computer located on network E would need to use the transmission facilities of either network C or network D. Because routers require time to examine the destination address in the IP header of a packet, check its routing table, and route the packet from the interface it was received on to another interface for transmission toward its ultimate destination, there is a delay or latency associated with each router through which a packet traverses. As a packet crosses more networks, it also crosses through additional routers, which cumulatively adds to the delay or latency encountered. In addition, as additional networks

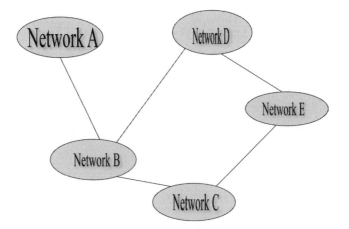

Figure 4.4 A computer user on network A needs to use the transmission facilities of third-party networks to reach a computer located on networks C, D, or E: peering and transit operations.

are traversed, the potential of encountering a network failure increases. As the number of networks traversed increases so does the delay and the probability that the data packet will not arrive at its intended destination.

Transit and peering operations

When two network operators interconnect their facilities to exchange data, they need to decide on the type of data exchange they are willing to support. If a network operator is willing to accept traffic from the other network that is destined for a different network, then the network operator is providing a transit facility to the other network. The network operator in effect has agreed to receive traffic from the other network and to pass it through its network onto the Internet regardless of its destination. Because Internet traffic is bidirectional, this means the network operator also agrees to receive traffic from the Internet and pass such traffic through its network.

If instead of agreeing to accept any traffic, let's assume both network operators are only willing to accept traffic from the other network that is destined for them. In this situation, the two networks are peering with one another. As long as the data from one network is destined to the other, it can pass through the peering connection while all other traffic will be blocked.

Because there is no free lunch in the field of communications, most organizations will charge a fee for providing a transit capability. That fee can vary, ranging from a fee structure based on the bandwidth of the connection to an amount per Mbyte or Gbyte of data that flows through one network from another network. A no-cost transit agreement usually occurs when a big network provider has interconnections to other similar sized

A. Peering without a peering point   B. Peering with a peering point

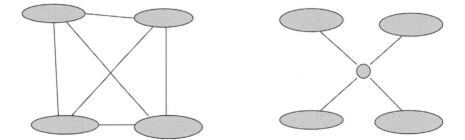

Figure 4.5 Understanding the value of a peering point.

providers. In comparison, a for-fee transit agreement is usually established when smaller network providers are only connected to one or a few other networks.

In comparison to transit operations, a peering point represents a location where many networks are interconnected to one another for exchanging traffic on a peering basis. To illustrate the advantages associated with peering points, consider the two groups of networks shown in Figure 4.5. In the left portion of Figure 4.5, peering is shown occurring among four networks without a common peering point. In this situation (n-1)/2 communications circuits are required for each network to be interconnected with every other network. In the example shown in the left portion of Figure 4.5 each network requires (4-1)/2 or 3/2 links. Because there are four networks to be interconnected, a total of 4 * 3/2 or 6 communications circuits are required to interconnect each network to every other network.

In the right hand portion of Figure 4.5, using a peering point is shown. In this example, the four networks are interconnected at a common location. Each network only requires one communications circuit for the interconnection, or a total of four circuits for the four networks.

The advantages associated with peering points involve both the simplicity of interconnections and the economics. For example, if the collection of four networks shown in Figure 4.5 were increased by just one, then the total number of communications circuits required to interconnect every network to each other without peering points would be 5 * (5-1)/2 or 10. Similarly, an increase in the number of networks to be interconnected to 6 would result in the need for 6 * (6-1)/2 or 15 communications circuits. In comparison, a peering point would only require each network to have one communications connection to the peering point, or a total of six communications connections to provide an interconnection capability for all six networks. Table 4.3 indicates

Table 4.3 Communications Circuits Required to
Provide Network Interconnections

Number of Networks to Be Interconnected	Without Using a Peering Point	Using a Peering Point
2	2	2
3	3	3
4	6	4
5	10	5
6	15	6
7	21	7
8	28	8
9	36	9
10	45	10

the number of communications circuits that would be required to interconnect networks with and without a peering point.

From an examination of the entries in Table 4.3, it's apparent that a
peering point can provide a significant reduction in communications links,
especially as the number of networks to be interconnected increases. Because
each communications link requires a router port and routers can only support a finite number of serial ports, after the maximum support level is
reached another router is required. Both routers and router ports represent
costs that need to be considered. In addition, when peering occurs without
a peering point the result can be a mesh network structure employed to
interconnect networks to one another. Under this situation, configuring routers becomes more complex then when a peering point is employed, because
the peering point only requires the routing of data over a single communications path for one network to obtain connectivity with all other networks.
Thus, the cost of equipment, personnel time and effort, and the complexity
of connecting networks without a peering point resulted in most network
connectivity now occurring via peering points.

Global structure of peering points

The concept associated with peering points is now on a global basis, with
locations throughout the United States, Europe, South America, Asia, and
Australia. Sometimes the term Metropolitan Area Exchange (MAE) is used,
while other terms used as synonyms include Network Access Point (NAP),
and Internet Exchange (IX). Regardless of the term, peering points are now
the primary method by which ISPs interconnect their separate networks to
enable traffic to flow throughout the Internet.

Representative peering points

Previously in this book, we looked at the statistics provided by one peering point in the United States and briefly discussed how peering points can now be encountered on a global basis. To obtain a better appreciation for the global nature of peering points, we will describe the European Internet Exchange (Euro-IX) and several European Exchange Points.

European Internet Exchange (Euro-IX)

Operators of European Internet Exchange Points set up the European Internet Exchange (Euro-IX). The goal of Euro-IX is to assist ISPs looking to peer at a European exchange point.

Table 4.4 lists European Exchange Points by country as of early 2005. There were over 90 exchange points operating in European countries. To obtain an appreciation of a European Exchange Point we will briefly examine three members of Euro-IX: the Vienna Internet exchange (VIX), the Belgium National exchange (BNIX), and the London Internet Exchange (LINX).

The Vienna Internet Exchange (VIX)

The Vienna Internet exchange (VIX), which is located at the Vienna University computer center, provides a peering point in the geographic center of Europe. A VIX member is required to be an ISP with its own Internet connectivity. In addition, other VIX members must not solely provide Internet connectivity, which in effect restricts membership to ISPs whose networks cover areas beyond central Europe.

Each VIX member is required to have its own Autonomous System Number (AS) as well as its own Internet registry, such as a Réseaux IP Européens (RIPE) or NCC subscription. A VIX member must also provide Internet access to its customers at the IP level. Thus, a content provider would not qualify for a VIX membership.

Membership costs

VIX services are provided on a not-for-profit basis. This means that its tariffs are set to recover the cost of operation and does not include any profit. Table 4.5 indicates the VIX tariff in effect during 2005, with pricing in euros (EUR). In early 2005, a euro was worth approximately $1.40, so readers can multiply EUR entries in the table by 40 percent to obtain an approximate charge.

In addition to the tariffs, there is a supplemental charge of 150 euros per month that is only applicable to the University of Vienna. This charge is for an ATM/OC-3 multi mode port on a LS1010 switch. Not considering the supplemental charge, users of VIX have three fees to consider — a setup fee, a VIX switch port fee, and a fee for housing equipment. The VIX switch port fee is based on the speed of the interconnection while the housing fee is

Table 4.4 Members of the Euro-IX by Country

Austria
Grazer Internet eXchange
VIX — Vienna Internet eXchange

Belgium
BNIX — Belgium National Internet eXchange
FREEBIX — Free Belgium Internet eXchange

Bulgaria
SIX

Croatia
CIX — Croatian Internet eXchange

Czech Republic
Neutral Internet eXchange

Cyprus
CYIX — Cyprus Internet eXchange

Denmark
DIX — Danish Internet eXchange

England
LINX — London Internet eXchange
LIPEX — London Internet Providers eXchange
LoNAP — London Network Access Point (now trailing multicast)
MaNAP — Manchester Network Access Point
UK6X
Xchangepoint — Multi-National

Estonia
TIX Tallinn Internet eXchange

Finland
FICIX Finnish Commercial Internet eXchange
Tampere Region Internet eXchange — TREX

France
EuroGix — A Peering Point
FNIX6 — eXchange in Paris
FreeIX — A free French eXchange
GNI — Grenoble Network Inititive
LYONIX — Lyon Internet eXchange
MAE — Paris
MAIX — Marseille Internet eXchange
MIXT — Mix Internet eXchange and Transit
PARIX — A Paris Internet eXchange
PIES — Paris Internet eXchange Service
PIX — Paris Internet eXchange

Table 4.4 (continued)   Members of the Euro-IX by Country

POUIX — Paris Operators for Universal Internet eXchange
SFINX — Service for French Internet eXchange

Germany
BECIX — Berlin Internet eXchange
BCIX — Berlin Commercial Internet Exchange
DE-CIX — Deutsche Commercial Internet eXchange
ECIX — European Commercial Internet eXchange (formally BLNX) Berlin
ECIX — Dusseldorf
HHCIX — Hamburg
INXS — Munich and Hamburg
Franap — Frankfurt Network Access Point
MAE — Frankfurt
KleyRex — Kleyer Rebstcker Internet eXchange (Frankfurt)
MANDA — Metropolitan Area Network Darmstadt
M-CIX — Munich Commercial Internet eXchange
N-IX — Nurnberger Internet eXchange
Work-IX Peering Point — Hamburg
Xchangepoint — Multi-National

Greece
AIX — Athens Internet eXchange

Hungary
BIX — Budapest Internet eXchange

Iceland
RIX — Reykjavik Internet eXchange

Ireland
INEX — Internet Neutral eXchange

Italy
MIXITA — Milan Internet eXchange
NaMeX — Nautilus Mediterranean Exchange Point (Rome)
TOPIX — Torino Piemonte IX
TIX — Tuscany Internet eXchange

Latvia
Latvian GIX

Luxembourg
LIX — Luxembourg Internet eXchange

Malta
MIX — Malta Internet eXchange

Netherlands
AMS-IX — Amsterdam Internet eXchange
GN-IX Groningen Internet eXchange
NDIX — A Dutch German Internet eXchange
NL-IX - NL — Internet eXchange

Table 4.4 (continued)   Members of the Euro-IX by Country

Norway
NIX — Norwegian Internet eXchange

Poland
WIX — Warsaw Internet eXchange

Portugal
GIGAPIX — Gigabit Portuguese Internet eXchange

Romania
BUHIX — Bucharest Internet eXchange
Ronix — Romanian Network for Internet eXchange

Russia
Chelyabinsk Peering Point Ural
MPIX
NSK — IX
RIPN Home Page (MSK—IX/M9—IX/SPB—IX)
Samara IX
Ural — IX

Scotland
World.IX — European Commercial IX (Edinburgh)
ScotIX — Scottish Internet Exchange

Slovenia
LIX — Ljubljana Internet eXchange

Slovak Republic
SIX — Slovak Internet eXchange

Spain
CATNIX — Barcelona
ESPANIX — Spain Internet Exchange
GALNIX — Galicia Internet eXchange
MAD-IX — Madrid Internet eXchange
Punto Neutro Espaol de Internet

Sweden
Linkoping Municiple Exchange
LIX — Lule Internet Exchange
NorrNod
NETNOD Internet eXchange
RIX -GH Gaveleborg Regional Internet Exchange
SOL-IX — Stockholm

Switzerland
CIXP — CERN eXchange for Central Europe
SWISSIX — Swiss Internet eXchange
TIX — Telehouseís Zurich eXchange

Table 4.4 (continued)   Members of the Euro-IX by Country

Turkey
TIX Turkish Internet eXchange (Temporarily Suspended: 16 Nov. 2004)
TIX — Turkish Information Center
TTNET — Turk Telekom Internet Backbone
TURNET — The Turkish Information eXchange

Ukraine
Central Ukrainian Internet eXchange
UA-IX — Ukrainian Internet Exchange
UTC-IX — Ukrtelecom Internet Exchange

Useful European Links
A Portal about European IXPs
APNIC SIG Link: Feb. 2003
Network Information Centers
DENIC — Deutsches Network Information Center
DKNIC — Danish Network Information Center
United Kingdom Network Information Center

Table 4.5 Vienna Internet Exchange Tariffs

	Fee in euros
Service Setup	*100/month*
VIX switch port	
10 Mbps (10Base T)	100/month
100 Mbps (100Base T)	300/month
1000 Gbps (1000Base SX)	1000/month
Housing for shelf in 19 inch rack	
Up to 3 height units	225/month
4 to 5 height units	375/month
6 to 9 height units	675/month

based on the amount of shelf space required for the equipment located at
the Vienna Internet exchange.

Belgian National Internet Exchange (BNIX)

Moving northwest from Vienna, Austria, we will turn to the Belgian National
Internet Exchange (BNIX). BNIX was established in 1995 by the Belgian
National Research Network (BELNET) and represents a peering point where

ISPs can exchange traffic with each other in Belgium. In addition to ISPs, BNIX is open to an alternative telecommunications operator and by 2005, telephone companies had introduced optical connections that provided BNIX with an abundant source of bandwidth.

Currently BNIX is constructed based on a distributed layer 2 switched medium consisting of Fast Ethernet (100 Mbps) and Gigabit Ethernet (1000 Mbps) switches connected to one another using 10-gigabit Ethernet technology. According to BNIX, this interconnection method provides a high speed, congestion-free interconnection facility, which enables participating ISPs to exchange data without experiencing any significant bottlenecks. More than 40 ISPs used the facilities of BNIX to obtain interconnectivity with one another in Belgium.

London Internet Exchange, Ltd. (LINX)

In concluding our brief tour of representative European exchange points, we will focus on the London Internet Exchange, Ltd. (LINX). LINX represents the largest exchange point in Europe and is a founding member of Euro-IX.

LINX represents a not-for-profit partnership between ISPs, providing a physical interconnection for its members to exchange Internet traffic through cooperative peering agreements. Candidates for becoming a LINX member include having an Autonomous System Number (ASN) and using the Border Gateway Protocol Version 4 (BGP4+) protocol for peering.

Membership costs

Although the LINX tariff has some similarities to the VIX tariffs, there are also some significant differences between the two fee schedules. Table 4.6 provides a summary of the LINX tariff in effect during early 2005. In examining Table 4.6, note that in addition to a setup or joining fee, LINX charges members a quarterly membership fee. While LINX bills subscribers similar to VIX for port and rack space, LINX unlike VIX also has a traffic charge, which for large data exchanges can add up to a considerable expense. While LINX is similar to VIX in that both operate as not-for-profit entities, LINX currently provides interconnections at up to 10 Gbps which is a considerably higher peering rate than presently provided by the VIX. Although a 10 Gbps, interconnection should minimize bottlenecks, in actuality potential bottlenecks depend on the traffic exchanged at a particular point in time. In concluding our discussion of peering points we will return to the use of the trace route program to examine peering point delays.

Peering point delays

In concluding our discussion of peering points, we will return to the Microsoft tracert program included in different versions of the Windows operating system. To use tracert, you need to open an MS-DOS window in

Table 4.6 The London Internet Exchange Price List

Service		Payment Schedule	GBP	EURO
Joining fee		Once	1000	1500
Membership fee		Quarterly	625	938
Port Fees				
100 Mbps		Monthly	175	263
	1 Gbps	Monthly	644	966
second	1 Gbps	Monthly	483	725
	10 Gbps	Monthly	2415	3625
Traffic Charge				
	Per Mbytes	Monthly	0.60	0.86
Rack Space				
	Per unit	Monthly	50	75

older versions of Windows, or what is now referred to as the Command Prompt window when using more modern versions of the Windows operating system, such as Windows 2000 and Windows XP. When using a more modern version of the Windows operating system, you can locate the Command Prompt menu entry by selecting Start > Programs > Accessories > Command Prompt.

Using tracert

Tracert lets you determine where bottlenecks are occurring when you encounter delays in accessing a server. Although most persons have the inclination to cite the server as the contributing factor when experiencing slow response time, it's quite possible the delay resides in the Internet.

To determine if the network represents most of the delay, you could first use the ping program built into Windows. Ping provides you with the round trip delay in milliseconds (ms) to a defined IP address or host name. If the round trip network delay appears to be reasonable then the delay can be attributable to the server. In comparison, if the round trip delay provided using the ping program is relatively lengthy, then the response time delay you are experiencing has a significant network component. When this situation arises, you can use the tracert program to determine where the delays in the network are occurring.

To illustrate tracert, let's assume you are accessing www.londontown.com, a Web site that provides a variety of tourist services for persons visiting London,

Figure 4.6 The home page of Londontown.com

England. Figure 4.6 illustrates the home page of www.londontown.com. Note that from this site you can search for a hotel or bed and breakfast, arrange for airport transfers, book sightseeing tours, and reserve theater tickets to the best shows in London. Because London is one of the most popular tourist destinations in the world, as you might expect www.londontown.com represents a popular Web site.

Because the response to page requests to Londontown.com can be relatively long during an approaching holiday, let's use the tracert program to examine the delays associated with the route to that Web site. Figure 4.7 illustrates the tracert program. In this example, this author traced the route to Londontown.com from accessing the Internet in Macon, Georgia.

Note in examining the entries shown in response to the tracert program, each line of output corresponds to a hop the data has to go through to reach its destination. The first hop represents the delay associated with the author's connection to the Internet. Because all three tries have a latency under 10 ms we can assume that the access line is not congested.

The second hop is the flow of data to a router located in Atlanta. The third hop is a router located in Atlanta. If you carefully read the router descriptions for the routers associated with hops 2 and 3, you will note that the router located at hop 2 is associated with bbnplanet while the router at hop 3 is associated with level 3 communications, a wholesale telecom carrier. Between hops 2 and 3, data flows from one carrier to another due to a peering arrangement. Because both hops 2 and 3 are located in Atlanta, propagation delay is minimal and by hop 3, two out of three computed delays are shown

```
Command Prompt                                                              _ □ ×

C:\WINNT\Profiles\Administrator>tracert www.londontown.com

Tracing route to www.londontown.com [213.161.77.151]
over a maximum of 30 hops:

  1    <10 ms    <10 ms    <10 ms    205.131.176.1
  2    <10 ms    <10 ms    10 ms     s4-0-8.hsa1.atl1.bbnplanet.net [4.24.209.73]
  3    <10 ms    <10 ms    10 ms     ge-6-2-0.bbr1.atlanta1.level3.net [64.159.3.69]
  4    10 ms     20 ms     10 ms     ge-0-3-0.bbr2.washington1.level3.net [64.159.0.229]
  5    20 ms     20 ms     10 ms     so-6-0-0.edge2.washington1.level3.net [64.159.3.62]
  6    20 ms     20 ms     10 ms     abovenet-level3-oc48.washington1.level3.net [4.68.127.50]

  7    20 ms     20 ms     10 ms     so-4-0-0.mpr2.iad2.us.above.net [64.125.30.122]
  8    20 ms     20 ms     20 ms     so-4-0-0.cr1.iad1.us.above.net [64.125.28.213]
  9    10 ms     20 ms     20 ms     so-1-0-0.cr1.dca2.us.above.net [64.125.28.125]
 10    90 ms     90 ms     90 ms     so-6-0-0.cr1.lhr3.uk.above.net [64.125.31.185]
 11    80 ms     81 ms     90 ms     pos0-0.er1a.lhr3.uk.above.net [208.184.231.150]
 12    90 ms     90 ms     90 ms     www.londontown.com [213.161.77.151]

Trace complete.

C:\WINNT\Profiles\Administrator>
```

Figure 4.7 Using tracert to observe the delays in reaching www.Londontown.Com

to be under 10 ms while the third delay is shown to be 10 ms. For all three times the delay is minimal.

For hops 3 through 5, traffic remains on the level 3 network, where data is routed to Washington, D.C. Between hops 6 and 7, traffic exits the level 3 network and enters AboveNet. AboveNet is an all-optical network backbone that interconnects data centers in the United States and Europe to include locations in northern Virginia in the United States and London in the United Kingdom. Thus, hops 7 through 9 correspond to the routing of data on the AboveNet network in the United States. Note that by hop 9 for two out of three time measurements the delay is 20 ms, with the third delay time shown as 10 ms. Again, these are reasonable delays.

Propagation delay

From hop 9 to hop 10, data flows from the AboveNet router located in the United States to that vendor's router located in the United Kingdom. Note that the time delay significantly increases from between 10 ms and 20 ms at hop 9 to 90 ms at hop 10. This delay results from placing data onto a trans-Atlantic fiber cable and includes the propagation delay associated with data crossing the Atlantic Ocean.

Once data arrives in England at hop 10, the delay associated with reaching the Londontown.com Web site is negligible. Thus, the primary delay in this situation results from an approximate 70 ms time required for data from a router located in the United States to reach a router located in the United Kingdom. This propagation delay represents 70/90 or approximately 78 percent of the total delay and could be avoided if the Web site operator established an agreement with a content delivery provider that resulted in the distribution of their server content onto a computer located in the United States. While the peering point delay shown in this example was vastly exceeded by the propagation delay, this is not always true. As indicated in this section, using the tracert program you can determine the location where network delays occur, in effect obtaining a visual indication of network performance.

4.3   Edge Operations

From our examination of Internet bottlenecks, it's apparent that the centralized model of server-based content can result in several delays. Those delays include the bandwidth limitations associated with the access lines connecting the browser user and server to the Internet, peering point interconnection data rates, traffic, router hops traversed from the browser user to the server, and propagation delays based on the distance between the two. By moving content from a centralized location to servers distributed to areas that better match access requirements, many of the previously mentioned delays are minimized. For example, let's assume that a multinational Japanese organization has its server content distributed to edge servers located in Europe, Africa, the United States, Australia, and China from a single server residing in Tokyo.

Without the distribution of server contents all user requests would have to traverse several networks, multiple router hops, and more than likely pass through multiple peering points. However, with content now distributed onto servers located around the globe, browser users avoid most, if not all, of the previously mentioned bottlenecks. Moving content closer to groups of potential browser users represents a method to eliminate potential peering point bottlenecks, reduces the number of router hops data must traverse, minimizes propagation delays, and may even allow data flow to remain on one ISP network. Because a centralized server acquires browser user requests similar to a funnel with a single entry point in the form of a Web server's access line, moving content to distributed edge servers removes a centralized Web site's network access constraint. Now we have a basic appreciation for the advantages associated with moving Internet content onto distributed servers commonly referred to as edge servers, let's turn to how edge server operations are performed.

Operation

There are several commercially available CDN operators, each employing a slightly different method of operation. Because Akamai Technologies can represent the leader in the field of CDN providers, we will focus on the CDN operation of this vendor. We will note their support for an emerging standard in the form of a markup language that provides the ability to easily distribute content onto edge servers.

The Akamai network

The Akamai network consists of approximately 15,000 servers. Those servers are distributed across the globe and are connected by approximately a thousand communications carriers in more than 60 countries. Today Akamai provides a CDN facility that is used by over 1200 of the world's leading electronic commerce organizations.

Type of content support

Through the year 2000, most CDN providers focused their efforts on delivering static content. While the delivery of such content was satisfactory for many organizations, the growth in electronic commerce and the development of tools to enhance Web pages with varying content increased the need to support both dynamic and personalized content. Today, several CDN providers including Akamai Technologies are capable of distributing the entire contents of a customer's Web site, which includes static and dynamic pages and varying embedded objects. To obtain an appreciation for the way content delivery operates, let's first examine how browser requests are commonly fulfilled by a centralized electronic commerce Web site.

Centralized Web site access

Let's assume a browser user wants to purchase the latest release of a movie on DVD. That person might access an electronic commerce site and search for a particular DVD, such as the Star Wars Trilogy DVD. When the browser user fills in a search line and clicks on a button, his or her search entry is forwarded from the Web server to an application server as illustrated in the right portion of Figure 4.8. The application server uses the string query to perform a database query. The application server assembles a page based on information provided by the database server, common page components as the site's logos, a navigation menu, and perhaps selected advertising based on the type of product queries. The browser user receives the assembled page that allows them to add the DVD to their shopping cart, select a suggested title generated by the application server, or perform another action.

   If we assume another browser user makes the same request, some of the same series of steps will have to be performed again. The browser user's request will flow to the Web site, which in turn will pass the search query to an application server. That server might then check its cache memory to determine if the requested page can be rapidly delivered. If a copy of the page is not in memory, the application server will have to recreate the page. Although retrieval of the page from cache memory can slightly enhance response time, pages must still flow from the centralized site back to different browser users, resulting in a variable delay based on the number of router

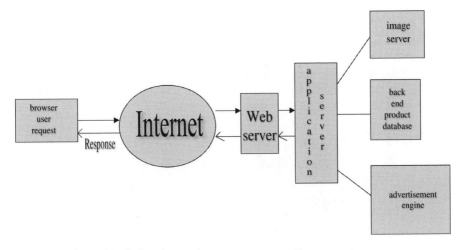

Figure 4.8 Centralized site electronic commerce operation.

hops and peering points that must be traversed, and the traffic activity on the Internet for a page to reach the requestor.

Edge server model

In comparison to the centralized Web site model, when Akamai edge servers are used, all initial Web page queries for a site that has contracted with the vendor flows to the vendor, while subsequent requests flow to an Akamai edge server. The edge server checks its internal cache to determine if the requested page was previously requested by another browser user and still resides in memory. Concerning cacheability of content, Akamai provides each Web site it supports with a metadata configuration file. The Web site manager uses that file to define how often specific pages can change.

For our previous DVD example, let's assume the electronic commerce site operator only changes their DVD pricing at most once per day. Then, the Web site manager would assign the DVD page a TTL value of one day. Thus, the first time a browser user requests the page it will be assembled by the Web site application server as illustrated in Figure 4.8. Since the page has a TTL value of one day, this page will be stored on Akamai edge servers for the same time, enabling subsequent requests for that page to be directly served to a browser user closer to the user than a centralized Web site. Figure 4.9 illustrates the data flow associated with the Akamai edge servers.

In examining the dataflow shown in Figure 4.9, note that although the Web page was created dynamically at the centralized Web site, the entire page can be stored on the Akamai network. Because the page was assembled for an individual user, there are no user-specific components, such as the personalization of a page via cookies that could prohibit the caching of the page.

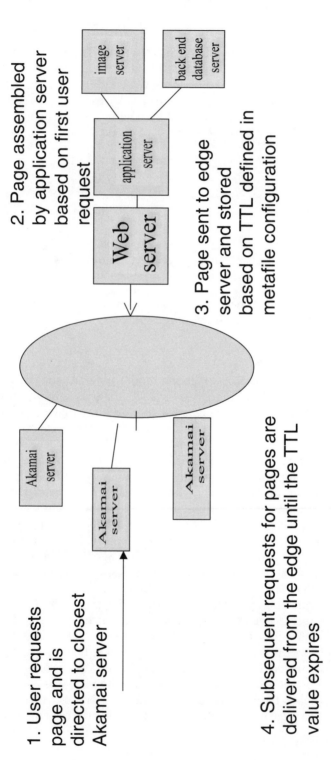

Figure 4.9 Dataflow using Akamai edge servers.

Limitation

The key limitation associated with the distribution of Web server content onto edge servers concerns the cookies and the agents for page personalization. Such sites as Yahoo!, MSN, and other portals use cookies to create a dynamic and personalized user experience. For example, consider Figure 4.10 and Figure 4.11. Figure 4.10 illustrates the Yahoo! home page prior to this author signing into Yahoo! to access his mail. In comparison, Figure 4.11 illustrates the Yahoo! home page after this author signed into Yahoo!. Note that in Figure 4.11 the page is slightly personalized, with Hi, gil_held shown in bold about one-third down the left portion of the page. Using a cookie to create a dynamic and personalized experience, Yahoo! also remembers that this author is signed in as he moves about the portal to check other features in addition to email.

Although the cookies enable Web page personalization, sites using cookies are normally considered to be non-cacheable. This means that the centralized Web site must maintain persistent connections to Akamai edge servers. Although the edge servers must then communicate with the centralized Web site, the original site only needs to have a finite number of connections to edge servers instead of tens of thousands or more connections to individual browser users. In spite of not being able to cache dynamic pages, the serving of uncacheable content via edge servers offers several advantages. Those

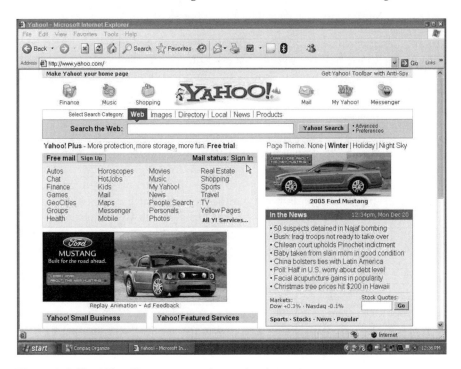

Figure 4.10 The Yahoo! home page prior to sign-in.

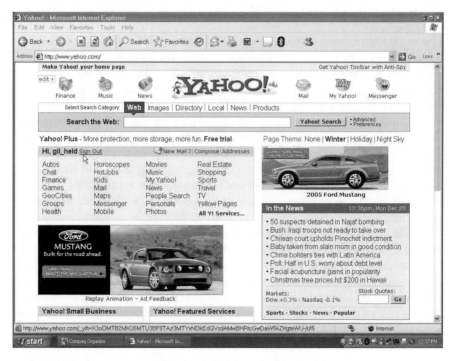

Figure 4.11 The Yahoo! home page after the author signed in.

advantages include the ability to offload CPU and memory from a centralized server to the ability of edge servers to respond faster to browser users
than serving requests from a central site. In addition, because edge servers
are located around the globe reliability of browser user access is increased.

Edge Side Includes (ESI)

Akamai Technologies in collaboration with application server and content
management organizations, developed a new markup language known as
Edge Side Includes (ESI). ESI represents a simple markup language used to
define Web page fragments for dynamic assembly at the edge. Using ESI,
files can be retrieved, which can be used to dynamically construct a Web
page in response to browser user requests. Each file can be controlled with
its own TTL value, which defines the time it will reside in cache memory,
enabling only small portions of a page to be retrieved from a central Web site
to build a full page for delivery to a browser user. Cookies could be retrieved
from the central Web site while static portions of a page might be stored in
cache memory on edge servers. This action enables an Akamai edge server
to assemble a Web page instead of the application server connected to a
centralized Web site. Because pages are assembled closer to the browser user,
this enables pages to be delivered faster. In addition, because more requests

are serviced on edge servers, this action reduces traffic to the centralized Web server.

Edge Side Includes support

ESI is the key to the ability to move dynamic content onto edge servers. Both the application server and Akamai edge servers must support the ESI language to enable applications to be deployed so browser users can obtain edge server support. The development of content using ESI begins at the centralized Web site with the development of templates and the creation of fragments. This is followed by the local assembly of pages and their placement in cache memory at the central Web site, their distribution to Akamai edge servers for remote assembly, and page caching. When a browser user request is directed to the central site, they can first retrieve information from cache at the central Web site, a location referred to as the edge of the data center. For subsequent requests, browser user requests are directed to an edge server for processing. Because the ESI language represents the mechanism by which application servers and edge servers communicate, let's obtain an overview of its capabilities.

ESI represents a simple markup language used to define Web page components for dynamic assembly and delivery of Web applications onto edge servers. ESI represents an open standard specification co-authored by application server and content management vendors. At the time this book was prepared, vendors supporting the ESI effort included Akamai Technologies, ATG, BEA Systems, Circadence, Digital Island, IBM, Interwoven, Oracle, Sun, and Vignette.

The primary benefit of ESI is it accelerates the delivery of dynamic Web-based applications. This markup language enables both cacheable and non-cacheable Web page fragment to be assembled and delivered at the edge of the Internet. To provide this capability, ESI is a markup language and specifies a protocol for transparent content management delivery. By providing the capability to assemble dynamic pages from fragments, it becomes possible to limit the retrieval of data from a centralized Web site to non-cacheable or expired fragments. This capability results in a lesser load on the centralized Web site, reducing potential congestion, and enhancing delivery of data to browser users.

The ESI markup language includes four key features: inclusion, conditional inclusion, environmental variables, and exception and error handling.

Inclusion and conditional inclusion

Inclusion provides the ability to retrieve and include files to construct a Web page, with up to three levels of recursion currently supported by the markup language. Each file can have its own configuration to include a specified TTL value, enabling Web pages to be tailored to site operator requirements. In comparison, conditional inclusion provides the ability to add files based on Boolean comparisons.

Environmental variables

A subset of standard Common Gateway Interface (CGI) environmental variables is currently supported by ESI. Those variables can be used both inside of ESI statements and outside of ESI blocks.

Exception and error handling

Similar to Hyper Text Markup Language (HTML), ESI provides the ability to specify alternate pages to be displayed in the event a central site Web page or document is not available. Under ESI, users could display a default page when certain events occur. ESI includes an explicit exception-handling statement set that enables different types of errors to generate different activities.

Language tags

Similar to HTML, the ESI specification defines a number of tags. Table 4.7 summarizes the function of seven key ESI tags.

The ESI template

The basic structure a content provider uses to create dynamic content in ESI is referred to as a template page. That page, which is illustrated in Figure 4.12, contains one or more HTML fragments that are assembled to construct the page. As indicated in Figure 4.12, the template is formed using such common elements as a vendor logo, navigation bars, similar canned static elements,

Table 4.7 ESI Tags

Tag	Function
<esi: include>	Include a separate cacheable fragment
<esi: choose>	Conditional execution under which a choice is made based on several alternatives, such as a cookie value
<esi: try>	Permits alternative processing to be specified in the event a request fails
<esi: vars>	Allows variable substitution for environmental variables
<esi: remove>	Specify alternative content to be removed by ESI but displayed by the browser if ESI processing is not performed
<esi... - ->	Specify content to be processed by ESI but hidden from the browser
<esi: inline>	Specify a separate cacheable fragment's body to be included in a template

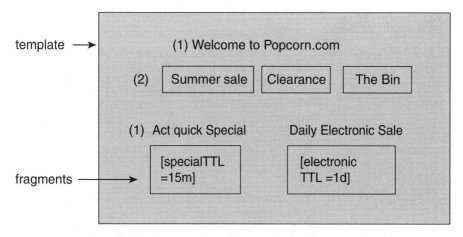

Figure 4.12 A sample ESI template page.

and dynamic fragments. The formed template represents a file associated with the URL a browser user requests. The file consists of HTML code that is marked up with ESI tags that informs the cache server or delivery network to retrieve and include the pre-defined HTML fragments, with the file constructed by combining HTML and ESI tags. In examining the ESI template page, shown in Figure 4.12, note that the Welcome logo (1) and the text (2) represent static content that can be permanently cached. The targeted advertisements (3 and 4) represent fragments that can only be stored in cache until their TTL values expire, after which edge servers must retrieve new fragments.

If we examine the sample ESI template shown in Figure 4.12, we can obtain an appreciation for how ESI considerably facilitates the flow of data. In this example, the Web page consists of a static boilerplate in the form of a vendor logo, page headings, and navigation bars that can be continuously cached on an edge server. The two fragments have TTL values of 15 minutes and 1 day, respectively. Thus, one fragment must be retrieved four times an hour from the central Web site, however, once retrieved, the fragment can be cached until the TTL value expires. The edge server only has to periodically update the contents of this fragment throughout the day. In comparison, the second fragment is only updated on a daily basis. An edge server only needs to retrieve one fragment every 15 minutes and the other fragment on a daily basis, to keep the Web page up-to-date.

Edge Side Includes for Java

In addition to ESI for HTML, an extension to ESI provides support for Java. Referred to as Edge Side Include for Java (JESI), it makes it easy to program Java Server Pages (JSPs). As a refresher, JSPs represent server-side software modules used to generate a user interface by linking dynamic content and static HTML through tags. Table 4.8 provides a summary of JESI tags including their function.

Table 4.8 JESI Tags

JESI Tag	Function
<jesi: include>	Used in a template to indicate to the ESI processor the manner by which fragments are assembled to form a page
<jesi: control>	Used to assign an attribute to templates and fragments
<jesi: template>	Used to contain the entire contents of a JSP container page within its body
<jesi: fragment>	Used to encapsulate individual container fragments within a JSP page
<jesi: codeblock>	Used to specify that a piece of code should be executed before any other fragments being executed
<jesi: invalidate>	Used to remove or execute selected objects cached in an ESI processor
<jesi: personalize>	Used to insert personalized content into a Web page where the content is placed in cookies that the ESI processor uses to insert into a page

Statistics

One of the major advantages associated with edge servers is the intelligence provided by some vendors in the form of statistical reports. For example, Akamai Site Wise reports provide information concerning which site pages are viewed, how long visitors remain on a site, the amount of time visitors spend on different pages, and similar information. Table 4.9 lists some of the Web site statistics provided by Akamai Site Wise reports and the potential utilization of such reports

As indicated in Table 4.9, such reports provide a tool for tailoring marketing and advertising resources and provide organizations with a window into the use of their Web content. Concerning Web content, information about the access and use of Web pages can be a valuable resource for tailoring content to reflect price sensitivity and product requirements of browser users that can turn page hits into shopping cart fulfillments. This in turn can result in increased revenues.

4.4 Summary

As indicated by our brief tour of ESI, this markup language represents the key to making edge operations practical. By providing the ability to subdivide Web pages into fragments, it becomes possible to cache dynamic portions of Web pages on a periodic basis, reducing the amount of traffic that has to flow between browser users and a centralized Web site.

Table 4.9 Akamai Site Wise Statistics and Potential Utilization

Statistic	Potential utilization
Most requested page	Move popular content to home page Eliminate minimally used content Leverage popularity of content
Routes from entry	Identify points of entry to target advertising
Popular routes	Determine if page organization is properly constructed for visitor access
Transactions by product	Determine online revenue drivers Test price changes and compare product sales
Shopping cart summary	Track fulfillment versus abandonment by product Contact customers who abandoned items with a special offer
Search engine summary	Determine which search engines refer customers to target advertising
First time vs returning customers	Determine value of different marketing programs Tailor content for repeat visitors Examine purchase patterns

chapter five

Caching and Load Balancing

In previous chapters in this book, our primary focus was on understanding the benefits of content delivery, reviewing the relationship between Web servers, application servers, and back-end database servers, and examining how a content delivery provider, such as Akamai, structures their network and uses a markup language to facilitate data delivery. Although our prior explanation of Content Delivery Network (CDN) operations was of sufficient detail to provide readers with a firm understanding of CDN operations, two key content delivery functions were glossed over. Those functions are caching and load balancing, both of which will be covered in more detail in this chapter.

The rationale for caching and load balancing being presented in the fifth chapter of this book is based on the structure of this book. The first three chapters provided a solid review of the rationale for content delivery and the interrelationship of Web requests and server responses, Chapter 4 focused on using a CDN service provider. If you use a CDN service provider, that provider will more than likely perform caching and load balancing transparently. However, if your organization has more than one Web server or if your organization decides that the enterprise should perform content delivery, then caching and load balancing needs to be considered. In this chapter, we will focus on both topics, examining how caching and load balancing operate and the advantages and disadvantages associated with each technology.

5.1 Caching

Caching represents a technique in which previously retrieved information is held in some type of storage to facilitate a subsequent request for the same information. Storage can include random access memory (RAM), disk, or a combination of the two. There are several types of caches where data can be temporarily stored, ranging from a user's browser to the server and other devices along the request/response path. To obtain an appreciation of caching, let's turn to the different types of caches that can be used to expedite the delivery of data, commencing with the browser cache.

Browser cache

Earlier in this book when we discussed the operation of browsers, we noted that a browser cache resulted in previously retrieved Web pages being stored on disk. Depending on the settings for your Web browser, your browser could check for a newer version of stored Web pages on every page retrieval request, every time you start Microsoft's Internet Explorer, automatically, or never. The selection of every page results in the browser checking whether a copy of the page to be viewed is cached every time you start Microsoft's Internet Explorer means that the browser checks to see if a copy of the page to be viewed was put in cache on the current day. If you configure Internet Explorer for one of the first three options, a request for a Web page will result in the browser comparing the parameters of the Web page, such as its created and modified date and file size to any previously stored page in cache. If the properties of the requested and stored page do not match, the browser will retrieve a new copy of the page. Obviously, if you selected the never option the browsers would not check the properties of the requested page. Instead, it would display the cached version of the page.

The primary purpose of a browser cache is to provide a more efficient method for retrieving Web pages. Instead of having to retrieve a previously retrieved Web page from the Internet, the page can be displayed from cache. This is more efficient because it minimizes latency or delay while reducing traffic flow on the Internet.

You can appreciate the usefulness of a browser cache when you click on the browser's back button or on a link to view a page you recently looked at. In such situations the use of the browser cache results in the near instantaneous display of a Web page. In comparison, if cache is not used, the delay to display a page can be as long as 20 or more seconds when the page contains a lot of images and is retrieved via dialup.

Other types of Web caches

In addition to browser caches there are several other types of caches whose operation directly affects the delivery of Web content. Those additional caches include proxy caches, gateway caches, and server caches.

Proxy caches

Web proxy caches operate very similar to browsers caches; however, instead of providing support for a single computer, they are designed to support hundreds to thousands of computers. A proxy server typically resides on the edge of an organization's network, usually behind the router that provides connectivity to the Internet. The proxy cache can operate as a standalone device or its functionality can be incorporated into another device, such as a router or firewall.

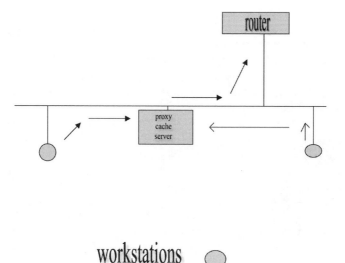

Figure 5.1 A proxy cache server supports a number of browser users.

Figure 5.1 illustrates the relationship between browser users, a stand-alone cache, and a router connected to the Internet. For browser users to effectively use the services of a proxy cache, they need to access the cache. The browser user requests have to be routed to the proxy. One way to accomplish this is to use your browser's proxy settings to tell the browser what services should be routed to the proxy. If you are using Internet Explorer, you would go to Tools> Internet Options > Connections and select the button labeled LAN Settings. This will display the Local Area Network (LAN) Settings dialog box as illustrated in the left portion of Figure 5.2. If you click on the Proxy Server box, you would be able to enter the address and port number of the proxy. As an alternative, you could click on the Advanced button, if your organization operates multiple proxy servers or if you wish to define exceptions to using a proxy server. The right portion of Figure 5.2 illustrates the Proxy Settings dialog box, which enables you to define multiple proxy servers and exceptions to using a proxy server.

In comparing browser caches to proxy caches, you can view the proxy cache as a type of shared browser cache because the proxy serves a large number of browser users. Similar to a browser cache, a proxy cache reduces both latency and network traffic. Because the proxy cache stores Web pages requested by many browser users, they are more efficient than individual or even a series of individual browser caches because the proxy cache stores pages previously accessed by a large group of users. This allows one user to have his or her request for a Web page fulfilled by transmitting a Web page previously cached because of the activity of another user who visited the location.

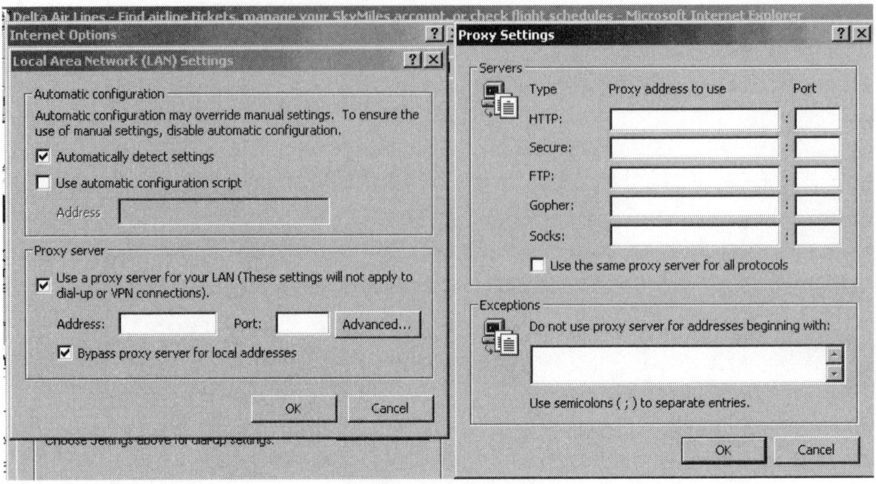

Figure 5.2 Internet Explorer, as well as other browsers, supports the use of proxy servers for one or more Web applications.

Gateway caches

A gateway cache can represent a reverse proxy cache because a gateway cache is typically installed by a Webmaster to make their site more scaleable. In comparison, a network manager installs proxy cache to conserve bandwidth. From a network perspective, the proxy cache resides in front of a network used by a group of browser users, while the gateway cache resides in front of a Web server used by distributed browser users whose access requests can originate from different networks.

While a gateway cache is commonly installed at a Web site, this device can also be distributed to other locations, with a load balancer used to route requests from a Web server to an individual gateway cache. When gateway caches are distributed across the Internet, you obtain a CDN capability.

Server caches

Server cache is a function of the operating system, the hardware platform, and the application program that provides a Web site capability. The operating system typically stores previously accessed files in RAM, with the number of files stored a function of available memory and the size of each file. In comparison, the application program, which provides a Web server capability, may allow you to cache a number of popularly requested Web pages in RAM, such as the home page that is retrieved when a browser user accessed your site.

Application caching

One example of application caching occurs with ASP.NET, a Microsoft server-side Web technology used to create Web pages. Essentially, ASP.NET treats every element in an ASP.NET page as an object, compiling the page into an intermediate language. Then, a Just In Time (JIT) compiler converts the intermediate code into native machine code that is executed on the host. Because the host directly executes the code, pages load faster than conventional Active Server Page (ASP) pages where embedded Visual Basic Scripting (VB Script) or JavaScript (Jscript) had to be continuously interpreted.

Under ASP.NET, frequently accessed pages can be cached via directives located at the top of each Active Server Page Framework (ASPX) file. Page developers can declare a specific ASPX page for caching by including the Output Cache directive at the top of the file. The following example illustrates the format of the Output Cache directive.

```
<%@Output Cache

    Duration="of seconds"

    Location= "Any|Client|Downstream|Server|None"

    Vary By Custom= "browser|customstring"

    Vary By Header="headers"

    Vary By Param="parametername" %>
```

The above coding is an example of enabling caching for a page declaratively. You can also enable caching programmatically in a page's code by manipulating the HttpCachePolicy object. Both methods work the same way for basic caching.

The first attribute, Duration, specifies how long in seconds to cache a Web page. Once a page is generated, ASP.NET will place it into cache. Then, until the duration is reached, subsequent requests for the same page will be served from cache. Once the specified duration is reached, the page is discarded. However, the next request for the page results in its generation and placement into cache, starting the process over again.

The second attribute, Location, enables you to specify where the cached Web page resides. The default setting of Location="any" caches the page on the client that originated the request, on the Web server that receives the request, or on any proxy servers located between the client and server that support HyperText Transfer Protocol (HTTP 1.1) caching. The remaining location attributes permit caching to occur at specific areas. For example, Location="Client" forces the page to be cached in the browser, Location="Server" results in the page being stored in the Web server cache, Location="Server And Client" uses the Web server or browser cache while Location="Downstream" results in the page stored anywhere other than the client browser.

The VaryBy... attribute can be used to cache different versions of the same page. Differences in Web pages can result from different client browsers, using different query strings or form-content parameters, and HTTP header values. For example, if your Web site provides several popular products whose descriptions are displayed via clients returning ProductId 27 and 54, in response to being queried, if you specify VaryByParam= "ProductId", ASP.NET processes the page and caches it twice, once for ProductID=27 and once for ProductID=54. In comparison, without this attribute only one version of the page would be cached. As indicated by this brief tour of ASP.NET caching, it provides a valuable technique to reuse previously performed processing used to create Web pages used in subsequent requests for the same data. Now we have an appreciation for the four common types of Web page caches, let's turn to how caches operate.

Cache operation

Regardless of the type of cache, each cache operates according to a set of rules, which defines when the cache services a Web page request. Some of the rules are set by an administrator, such as the browser operator or the proxy administrator. Other rules are set by the protocol, such as HTTP 1.0 and HTTP 1.1. In general, a cache examines traffic flow content and creates a cache entry based on the rules it follows. Table 5.1 lists some of the more common rules followed by a cache. As we review HTML META Tags and HTTP headers, we will note how the rules listed in Table 5.1 can be applied to the traffic that the cache examines.

In examining the entries in Table 5.1, note that the term fresh means that data is available immediately from cache. An age-controlling header or a timer provides a way to determine if data is within the fresh period.

Table 5.1 Common Cache Rules

• Examine response headers to determine if data should be cached. If the header indicates data should not be cached, it is not cached.
• If the response does not include a validator such as a Last-Modified header, consider data to be uncacheable.
• If the request is secure or authenticated do not cache data.
• Consider cached data to be fresh if:
– Its age is within the fresh period.
– A browser cache previously viewed the data and has been set to check once a session.
– A proxy cache recently viewed the data and it was modified a relatively long time long ago.
• If data is stale, the origin server will be asked to validate it or inform the cache if the copy is still valid.

Cache control methods

The most common method used to control the way caching operates is through HTML META tags and HTTP headers. As a review for some readers, META tags are HTML tags that provide information describing the content of a Web page. However, unlike HTML tags that display information, data within a META tag is not displayed, a term referred to as non-rendered.

META tags

META tags are optional, and many Web page developers do not use such tags. However, because the META tags are used by search engines to enable them to more accurately list information about a site in their indexes the use of this type of HTML tag grew in popularity.

Types of META tags

Two basic types of META tags — META HTTP-EQUIV and META tags have a NAME attribute. META HTTP-EQUIV tags are optional and are the equivalent of HTTP headers. Similar to normal headers, META HTTP-EQUIV tags can be used to control or direct the actions of Web browsers in a way similar to normal headers. In fact, some servers automatically translate META HTTP-EQIV tags into HTTP headers to enable Web browsers to view the tags as normal headers. Other Web server application programs employ a separate text file that contains META data.

The second type of META tag, which is also the more popular type of tag, is a META tag with a NAME attribute. META tags with a NAME attribute are used for META types that do not correspond to normal HTTP headers, enabling specialized information to be incorporated into a Web page.

Style and format

META tags must appear in the HEAD of an HTML document and are normally inserted at the top of a document, usually after the <TITLE> element. The format of a META tag is shown below:

```
<META name = "string1" content = "string2">
```

Note the META tag format does not need a </META> at the end of the tag. Table 5.2 provides an alphabetically ordered list of some of the major types of META tags and a brief description of their potential utilization. To obtain a better appreciation for META tags, let's examine a few examples. Suppose the content of a Web page should expire on July 19, 2005 at 2 P.M. Then, we would use the Expires META tag as indicated below:

```
<META name = "Expires" content = "Tue, 19 JLY 2005
14:00:00 GMT">
```

Table 5.2 Basic Types of META Tags

Tag	Description
Abstract	Provides a one line overview of a Web page.
Author	Declares the author of a document.
Copyright	Defines any copyright statements you wish to disclose about your document.
Description	Provides a general description about the contents of a Web page.
Distribution	Defines the degree of distribution of the Web page (global, local or internal).
Expire	Defines the expiration date and time of the document being indexed.
Keywords	Provides a list of keywords that defines the Web page. Used by search engines to index sites.
Language	Defines the language used on a Web page.
Refresh	Defines the number of seconds prior to refreshing or redirecting a Web page.
Resource-type	Defines the type of resource for indexing, which is limited to document.
Revisit	Defines how often a search engine should come to a Web site for reindexing.
Robots	Permits a Web site to define which pages should be indexed.

It should be noted that the Expires META tag uses dates that conform to RFC `1123.

For a second example of the use of META tags lets turn our attention to the Refresh META tag. The use of this tag provides a mechanism to redirect or refresh users to another Web page after a delay of a specified number of seconds occurs.

Because the Refresh META tag is used within an HTTP-EQUIV tag, let's examine the format of the HTTP-EQUIV tag below:

```
<META HTTP-EQUIV= "varname" content="data"
```

Note that the HTTP-EQUIV tag binds the variable name (varname) to an HTTP header field. When the varname is Refresh, the HTTP-EQUIV tag can be used in the HEAD section of an index.html file to redirect a browser user to another location. For example, to redirect a browser user to www.popcorn.com after a 5-second delay, you could code the following tag:

```
<META HTTP-EQUIV="Refresh" content="5;url=www.
popcorn.com">
```

Because we are discussing caching, we will conclude our brief review of META tags with the tag used to inform browsers and other products not to cache a particular Web page. That tag is the pragma tag, whose use is shown below.

```
<META HTTP-EQUIV="pragma' content='no-cache'>
```

Although META tags are relatively easy to use, they may not be effective because they are only honored by browser caches, which read HTML code. META tags may not be honored by proxy caches because they very rarely read HTML code. In comparison, true HTTP headers can provide you with a significant amount of control over how both browser and proxy caches handle Web pages. Let's turn to HTTP headers and how they can be used to control caching.

HTTP headers

In examining HTTP headers, we will primarily focus on cache control headers.

Overview

Currently there are approximately 50 HTTP headers defined in the HTTP 1.1 protocol that can be subdivided into four categories — entity, general, request, and response. An entity header contains information about an entity body or resource while a general header can be used in both request and response messages. A request header is included in messages sent from a browser to a server, while a response header is included in the server's response to a request.

HTTP headers in response messages are sent by servers prior to HTML code. The HTTP header data are only viewed by a browser and are not displayed by the browser, thus they are transparent to the user.

Table 5.3 illustrates an example of an HTTP response header transported under HTTP 1.1. Note that the Date header is used to specify the date and time the message originated and represents a general type of header.

Table 5.3 A Typical HTTP 1.1 Response Header

```
HTTP/1.1 200 ok
Date: Tue, 19 Jly 2005 14:21:00 GMT
Server: CERN/3.1
Cache-Control: max-age=7200, must-revalidate
Expires: Tue, 19 Jly 2005 16:21:00 GMT
Last-Modified: Mon, 18 Jly 2005 12:00:00 GMT
ETag: "4f75-316-3456abbc"
Content-Length: 2048
Content-Type: text/html
```

The server HTTP header represents a response type header that provides information about the software used by the server to respond to the request. Because revealing a specific software version can provide hackers with the ability to match vulnerabilities against a server, many times only the basic information is provided by this header.

Of particular interest to us is the cache-control header, because this header specifies directives that must be obeyed by all caching mechanisms along the request/response path. In the remainder of this section, we will focus on this header. However, we first need to discuss the Expires HTTP header, because it controls cache freshness.

Expires header

The Expires HTTP header provides a way that informs all caches on the request/response path of the freshness of data. Because Expires headers are supported by most caches, it also provides a way to control the operation of all caches along a path. Once the time specified in the Expires header is reached, each cache must then check back with the origin server to determine if data changed.

Expires response headers can be set by Web servers in several ways. Web servers can set an absolute expiration time, a time based on the last time a client accessed the data, or a time based on the last time the data changed. Because many parts of a Web page contain static or relatively static information in the form of navigation bars, logos, and buttons, such data can become cacheable by setting a relatively long expiry time.

As indicated in Table 5.3, the only valid value in an Expires header is a date in Greenwich Mean Time (GMT). Although the Expires HTTP header provides a basic way to control caches, to be effective each cache must have a clock synchronized with the Web server. Otherwise, caches could incorrectly consider stale content as being fresh.

Cache-control header

The cache-control general header field within the HTTP 1.1 header specifies how all caching mechanisms along the request/response path should operate. The cache-control header includes one or more request and response directives that specify the way caching mechanisms operate and typically override default-caching settings. Our discussion of cache-control is mainly applicable to HTTP 1.1, because HTTP 1.0 caches may not implement cache-control.

Table 5.4 indicates the cache-control header format and directives available under HTTP 1.1 for the cache-control header. Note that cache directives are unidirectional because the presence of a directive in a request does not mean that the same directive has to be included in the response. Also, note that cache directives must always be passed through devices along the

Table 5.4 Cache-Control Header Format and Directives

```
cache-control = 'cache-control"  ";"  1#cache-directive
cache-directive =cache-request-directive  |  cache-response-
directive
cache-request-directive =
            "no-cache"
          | "no-store"
          | "max-age"  "-"  seconds
          | "max-stale"  "="  seconds
          | "min-fresh"  "{="  seconds
          | "no-transform"
          | "only-if-cached"
          | cache-extension
cache-response-directive =
            "public"
          | "private"  ["="  <"> 1#field-name <"]
          | "no-cache"  [  "="  <"> 1#field-name <"]
          | "no-store"
          | "no-transform"
          | "must-revalidate"
          | "proxy-revalidate"
          | "max-age"  "="  seconds
          | "s-maxage"  "="
          | cache-extension
```

request/response path to the destination even if an intermediate device, such as a proxy cache, operates on a directive. This is because a directive can be applicable to several types of caches along the request/response path and currently there is no method to specify a cache directive for a specific type of cache.

Cache-control can be subdivided into request and response headers. Each cache-control header can include one or more directives that define how cache-control should operate. Cache-control requests were supported under HTTP 1.0, while HTTP 1.1 introduced cache-control response headers, which provide Web sites with more control over their content. See the following example.

```
Cache-control: max-age=7200, must-revalidate
```

Included in an HTTP response, this header informs all caches that the content that follows is considered fresh for two hours (7200 seconds). In addition, the must-validate directive informs each cache along the request/ response path that they must comply with any freshness information associated with the content.

Directive application

Prior to describing the function of the directives associated with the cache-control header in detail a few words are in order concerning the application of a directive. When a directive appears without any 1#field-name parameter, the directive then is applicable to the entire request or response. If a directive appears with a 1#field-name parameter, it is then applicable only to the named field or fields and not to the remaining request or response. Now that we have an appreciation for the applicability of directives let's turn our attention to the function of the specific directives that can be included in a cache-control header. In doing so we will first examine the cache-request directives listed in the previously referenced table.

Cache-request directives

In examining the entries in Table 5.4, you will note that there are seven distinct cache-request directives and a mechanism that permits the cache-control header to be extended. The mechanism that permits the cache-control header to be extended occurs by assigning a token or quoted string to the cache-extension. As we review the operation of each cache-request directive, we will also discuss its use as a cache-response directive.

The no-cache directive

The purpose of the no-cache directive is to force caches along the request/response path to submit a request to the origin server for validation prior to releasing a cached copy of data. This function can be used to maintain freshness without giving up the benefits of caching. In addition, this directive can be used along with the public directive to assure that authentication is respected.

    If the no-cache directive does not include a field-name, this forces a cache to use the response to satisfy a subsequent request without the successful revalidation with the server. If the no-cache directive specifies at least one field-name, then a cache can use the response to satisfy a subsequent request. In this situation, the specified field-name(s) are not sent in the response to a subsequent request without a successful revalidation of the origin server, enabling the server to prevent using certain HTTP header fields in a response while enabling the caching of the rest of the response.

The no-store directive

In comparison to the no-cache directive, the no-store directive is relatively simple. It instructs caches not to keep a copy of data under any condition.

The max-age directive

The purpose of the max-age directive is to enable the client to indicate it is willing to accept a response whose age is less than or equal to the specified time in seconds. The max-age directive specifies the maximum amount of time data will be considered fresh. Unless a max-stale directive is included, the client will not accept a stale response.

The max-stale directive

The purpose of the max-stale directive is to enable a client to be willing to accept a response that exceeds its expiration time. If a max-stale directive is included, the client is able to accept a response that exceeds its expiration time by the number of seconds specified in the max-stale directive. By using a max-stale directive without a value, the client is willing to accept all stale responses.

The min-fresh directive

The purpose of the min-fresh directive is to enable a client to accept a response whose freshness is equal to or greater than its current age, plus the specified time in seconds. Using this directive enables a client to specify it wants a response that will be fresh for at least the number of seconds specified in the min-fresh directive.

The no-transform directive

Because the implementers of proxy and other types of caches found it useful to convert certain types of data, such as images to reduce storage space, it is possible for a transformation to cause potential problems. For example, transforming an x-ray stored as a lossless image into a Joint Photographic Experts Group (JPEG) image reduces storage but could result in losing important medical information. A no-transform directive is used to prevent this situation from occurring, because it informs caches on the request/ response path not to perform any transformation of data.

The only-if-cached directive

The purpose of the only-if-cached directive is to enable a client to cache only those responses it has stored and not to reload or revalidate data with an origin server. This directive is commonly used under poor network connectivity conditions, and it informs a client cache to either respond using a cached entry applicable to the request or with a Gateway Timeout status.

Cache control extensions

The purpose of cache control extensions is to enable the cache-control header field to be extended. There are two types of cache control extensions — informational and behavioral. An informational extension does not require a change in cache and does not change the operation of other directives. In comparison, behavioral extensions operate as modifiers to existing cache directives. When a cache control extension is specified, applications that do not understand the extension will default to complying with the standard directive. In comparison, if the extension directive is supported, this new directive will modify the operation of the standard directive. Now we have an appreciation for cache-request directives, let's turn to cache-response directives.

Cache-response directives

As indicated in Table 5.4, there are nine cache-response directives and a mechanism to extend the cache control header. Because we previously discussed several cache-request directives that are also applicable to cache-responses, we'll focus on cache-response directives that are not applicable for use on a request path. However, as we review the directives, we will briefly note when the directive functions in a similar manner to its use in a request header.

The public directive

The purpose of the public directive is to make authenticated responses cacheable by any cache. This cacheability holds even if the data would normally be non-cacheable.

The private directive

The purpose of the private directive is to indicate that all or a portion of a response message is intended for a single user and must not be cached by any shared cache. However, a private non-shared cache can cache the response. An origin server can use this directive to indicate that specified portions of a response are only applicable for a single user.

The no-cache directive

The no-cache directive functions in the same manner as a cache-request directive. It forces caches to submit a request to the origin server prior to releasing cached data.

The no-store directive

Similar to the no-cache directive, the no-store directive functions in the same manner as a cache-request directive. It instructs caches not to keep a copy of data in cache.

The no-transform directive

The no-transform directive functions in the same manner as a cache-request directive. This directive does not allow caches to transform data.

The must-revalidate directive

The purpose of a must-revalidate directive is to ensure a cache does not use an entry after it becomes stale. A cache will first revalidate an entry prior to responding to a request.

The proxy-revalidate directive

The proxy-revalidate directive is similar to the must-revalidate directive. However, this directive does not apply to non-shared user agent caches.

One common application for the proxy-revalidate directive is on a response to an authenticated request. Because the request was previously authenticated, the proxy-revalidate directive enables a user cache to store and later return the response without having to revalidate it.

The max-age directive

Similar to its use in the cache-request header, the max-age directive specifies the amount of time data will be considered fresh.

The s-maxage directive

The s-maxage directive is similar to the max-age directive. However, this directive has one significant difference in that it is only applicable to shared caches, such as a proxy cache.

Cache-extension

The cache-extension directive is similar to how it is used in a cache-request header. It enables the header to be extended.

Viewing HTTP headers

There are several methods available for viewing the full headers contained in HTTP. You can manually connect to a Web server using a Telnet client activated using port 80 by entering the command to open www.xyz.com:80. Once you connect to a particular site, you can use the GET command to request the representation. For example, if you want to view the headers for http:/www.xyz.com/index.html, you would first connect to www.xyz.com on port 80. Then, you would type:

```
GET/index.html HTTP/1.1 [return]

Host:www.xyz.com [return] [return]
```

You press the Return key once to display each line in the header. Unfortunately, due to security concerns, most Web site operators do not support Telnet. A good alternative is to use an HTTP header viewer.

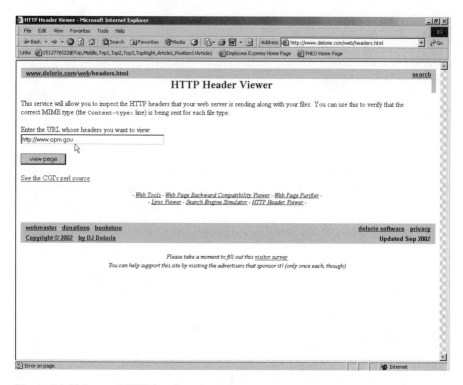

Figure 5.3 Using an HTTP header viewer.

One interesting HTTP header viewer can be used at www.delone.com/ web/headers.html. Figure 5.3 illustrates the HTTP header viewer screen into which this author entered a United States government URL. Figure 5.4 illustrates the HTTP headers for that Web site.

In examining the headers for the www.opm.gov Web site, we can note that the site is using Version 5.0 of Microsoft's server software. Because we are concerned with cache control, we will not discuss the other headers. Instead, let's turn to the private cache-control directive. The responses from this site are intended for a single user and must not be cached by any shared cache, although a private non-shared cache can cache the response.

In concluding our use of an HTTP header viewer, let's view the headers of the *New York Times*. Figure 5.5 shows the headers for www.nytimes.com. In examining Figure 5.5, let's focus on cache control. Note that the *New York Times* Web site uses a no-cache directive in its response, forcing caches to submit a request to the *New York Times* Web site server prior to releasing the cached data. As an educated guess, the reason behind using the no-cache directive is probably because the *New York Times* changes its page composition quite frequently. In fact, the right corner of its home page typically includes a Markets section showing stock averages and interest rates, which is updated every few minutes when the financial markets are open.

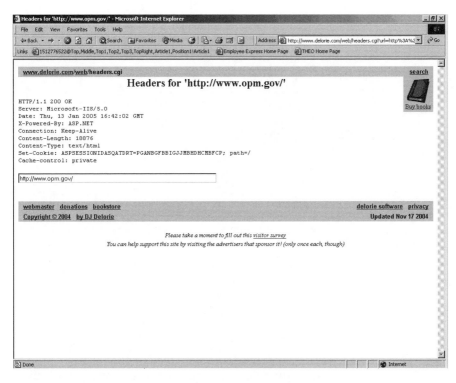

Figure 5.4 Viewing HTTP headers for www.opm.gov

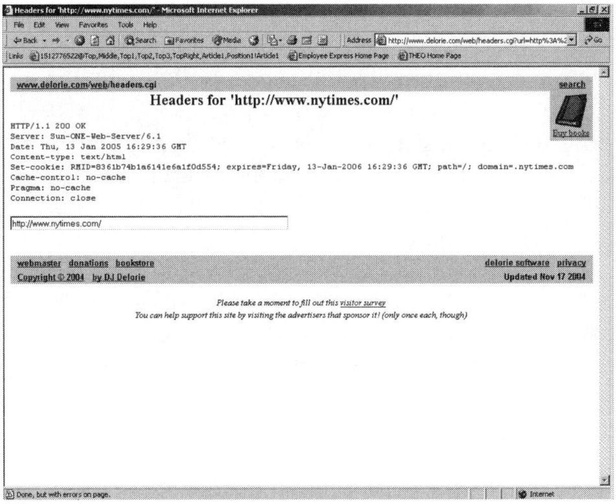

Figure 5.5 Viewing the HTTP headers for www.nytimes.com

Table 5.5 Techniques for Enhancing Cacheability
of Data

> - Minimize use of SSL/TLS
> - Use URLs consistently
> - Use a common set of images
> - Store images and pages that infrequently change
> - Make caches recognize scheduled, updated pages
> - Do not unnecessarily change files
> - Only use cookies when necessary

Considering authentication

In an electronic commerce environment, many Web pages are protected with HTTP authentication. Pages protected with HTTP authentication are considered private and are not kept by proxies and other shared caches. Some Web sites that want these pages to be cached can do so with the public directive. If a Web site wants those pages to be both cacheable and authenticated for each user, you would use both public and no-cache directives as shown below:

```
Cache-control: public, no-cache
```

The cache-control directives inform each cache that it must submit client authentication information to the origin server prior to releasing data from the cache.

Enhancing cacheability

In conclusion, let's focus on a core series of techniques that can be employed by Web site operators to enhance the cacheability of their pages. Table 5.5 lists eight techniques you can consider to improve the cacheability of data. Although some techniques may be more obvious than others may, let's discuss each in the order they appear in the referenced table.

Minimize use of SSL/TLS

When Secure Socket Layer/Transport Layer Security protocol (SSL/TLS) is used, pages flowing between the client and server are encrypted. Because encrypted pages are not stored by shared caches, reducing the use of SSL/TSL pages enhances the overall cacheability of Web site data.

Use URLs consistently

Using the same URL can make your site cache-friendly, enabling a larger percentage of pages to be cached. This is especially true when your organization provides the same content on different pages to different users.

Use a common set of images

If your Web site uses a common set of images and your pages consistently reference them, caching can become more frequent.

Store images and pages that change infrequently

If you carefully examine your Web site content, you will more than likely note many images and Web pages that are either static or change infrequently. By using a cache-control max-age directive with a relatively large value, you can make caches take advantage of infrequently changing or static data.

Make caches recognize scheduled updates

Certain pages, such as the *New York Times* home page, regularly change when financial markets are open. You can make caches recognize the scheduled change of these pages by specifying an appropriate max-age or expiration time in the HTTP header.

Do not unnecessarily change files

One of the problems associated with caching is its relatively easy for a Web site to associate a large number of files with false Last-Modified dates. For example, when updating a site by copying every file, instead of the files that changed, each file will appear to be recently modified and as a result, adversely effect caching. Web site operators performing backup and restores or other site update operations should restrict file updates to those files that changed.

Only use cookies when necessary

The purpose of a cookie is to identify the user's prior action. When a proxy server caches cookies, they provide no additional benefit than when cached by a browser. This means that more effective caching can occur if cookies are not used or if their use is limited to dynamic pages.

Minimize the use of POST

While information sent in a query via POST can be stored by caches, the reverse is not always true. Because most caches do not maintain responses to POST, minimizing the use of POST makes your data more cacheable.

5.2 *Load Balancing*

In comparison to caching that can be performed at many locations on the request/response path, load balancing is usually performed at the location where servers reside. In this section, we will examine the rationale for load balancing, the different algorithms that can be used to distribute HTTP requests among two or more servers, and the different types of load balancing available, including their advantages and disadvantages. In addition, we will examine the configuration and operation of a relatively free method of load balancing that occurs through the use of Domain Name System (DNS).

Rationale

Web servers are similar to other computers in that they have a set amount of processing power and RAM memory. Processing power, RAM memory, and the speed of the server's network connection limits the number of pages per unit time that can be served to clients. When an organization has only one Web server, its ability to respond to all incoming requests may be degraded as traffic increases. Clients will notice degraded server performance as requested pages load slowly or they experience timeouts and fail to connect to the server.

In an attempt to alleviate this problem, Web site operators have two options. First, they can upgrade their facility by adding multiple processors if the platform supports additional hardware. Next, they may be able to increase RAM memory, allowing more pages to be cached. However, an increase in traffic to the Web site can lead to the point where the continued upgrade of server hardware is no longer cost effective. Now, one or more servers need to be added to enable the load to be shared among a group of servers. The distribution of the load to be shared among two or more servers is referred to as load balancing

Load balancing techniques

Several techniques can be used to obtain a load balancing capability. Load balancing can occur through software, hardware, or a combination of software and hardware. In this section we will briefly review several types of load balancing techniques prior to examining them in detail.

DNS load balancing

Perhaps the easiest technique used to implement load balancing is through the modification of DNS entries. This technique, which is referred to as DNS load balancing, requires a site to have two or more servers, each running the same software. The address of each server is stored in the site's DNS under one name. The DNS server issues those addresses in a round-robin manner. The first client to access the site obtains the address of the first Web server the second client obtains the address of the second server, and so on. Later in this section, we will discuss DNS load balancing in more detail.

Internet Protocol address mapping

A second load balancing technique involves mapping the name of a Web site to a single IP address. That IP address represents the address of a computer or network appliance that intercepts HTTP requests and distributes them among multiple Web servers. This type of load balancing can occur using both hardware and software. Although using a load balancing appliance is more expensive than DNS modifications, it enables a more even load balancing to be achieved because a load balancing appliance can periodically check to see if each server it supports is operational and, if not, adjusts its serving mechanism. In comparison, DNS cannot check for the availability of servers and would periodically forward client requests to an inoperative server. This explains why when you attempt to access a Web site, you may first receive an error message when a subsequent access request results in the retrieval of the desired Web page. This is because the second request results in the DNS server returning the IP address of a different server on the second request. When multiple requests are required to access a Web server there is a high probability that the site uses DNS load balancing and at least one server is offline.

Virtual Internet Protocol addressing

A third load balancing technique involves using software to configure multiple servers with the same IP address, a technique referred to as virtual IP addressing. Under this technique, multiple servers can respond to requests for one IP address. For example, suppose you have three Windows 2003 servers, assigned IP addresses 198.78.64.1, 198.78.64.2, and 198.78.64.3, respectively. You could use virtual IP addressing software to configure all three servers to use the common virtual IP address of 198.78.64.5. You would then designate one server as the scheduler, which would receive all inbound traffic and route requests for Web content to the other two servers based on the load balancing parameters you set. Under virtual IP addressing, the failure of the scheduler server would be compensated for by assigning backup scheduling duties to another server.

Load balancing methods

Several load balancing methods can be used to distribute HTTP requests among two or more servers. Table 5.6 lists three of the more popular methods of load balancing algorithms in use.

Table 5.6 Types of Load Balancing Algorithms

• Random allocation
• Round-robin
• Weighted round-robin

Random allocation

Under the random allocation method of load balancing HTTP, requests are assigned to servers in a random manner. Although the random method of allocation is easy to implement, it's possible for one server to periodically be assigned more requests than another server although on average all servers over time should have the same load.

Round-robin allocation

A second method that can be used to assign HTTP requests to multiple servers is on a rotating or round-robin allocation method. Typically, the first request is allocated to a server selected randomly from a group of servers so not all initial requests are assigned to the same server. For subsequent requests, a circular order is used. Although a round-robin allocation method divides requests equally among available servers, it does not consider server processing capabilities. If one server has twice the processing capability of another server, a round-robin allocation method would result in the more powerful server having half the load of the less powerful server.

Weighted round-robin allocation

The weighted round-robin allocation method was developed in response to the processing differences between servers. Under the weighted round-robin load allocation method, you can assign a weight to each server within a group. If one server is capable of processing twice the load of a second server, the more powerful server would be assigned a weight of two while the less powerful server would be assigned a weight of one. Using the weight information, the more powerful server would be assigned two HTTP requests for each request assigned to the less powerful server. Although using a weighted round-robin allocation method better matches requests to server processing capability, it does not consider the processing times required for each request. In reality, considering the processing times required for each request would be extremely difficult because it would require every possible request to be executed on each server in the group and the processing requirements would have to be placed in a database on the load balancer.

Hardware versus software

There are several differences between hardware and software load balances that need to be considered prior to selecting a load balancing platform. A hardware load balancer is usually considerably more expensive than its software-based cousin. Although the hardware-based load balancer can provide a higher servicing capability, they may lack some of the configuration options available with software-based load balancers.

Today, a majority of load balances are software based. In fact, some Web server and application server software packages include a software load balancing module. Because DNS load balancing, one of the most popular load balancing methods, is software based and requires no additional hardware, we will conclude this section with a discussion of its operation.

DNS load balancing

As a review, the purpose of the DNS is to respond to domain name look-up requests transmitted by clients. In response to such requests, the DNS returns the IP address that corresponds to the requested domain name.

Under DNS load sharing, several IP addresses are associated with a single host name. When a request flows to the DNS server to resolve a domain name, it responds with one of the IP addresses that are served up in a round-robin or load-sharing manner. Serving up IP addresses in a round-robin manner results in a form of load balancing.

DNS load sharing methods

Two basic methods of DNS-based load sharing include CNAMES and A records.

Using CNAMES

One of the most common implementations of DNS is the Berkeley Internet Name Domain (BIND). Depending on which version of BIND is used, you can implement load balancing using multiple CNAMES or multiple A records.

Under BIND 4, support is provided for multiple CNAMES. Let's assume your organization has three Web servers configured with IP addresses 198.78.46.1, 198.78.46.2, and 198.78.46.3. You would add the servers to your DNS with Address (A Names) records as shown below:

```
Web1  IN  A  198.78.46.1

Web2  IN  A  198.78.46.2

Web3  IN  A  198.78.46.3
```

Note that the Web server names (Web1, Web2, and Web3) can be set to any name you want but need to match canonical names added to resolve www.domain.com to one of the servers through the following entries:

```
www  IN  CNAME  Web1.domain.com

     IN  CNAME  Web2.domain.com

     IN  CNAME  Web3.domain.com
```

Based on the preceding, the DNS server will resolve the www.domain. com to one of the listed servers in a rotated manner, spreading requests over the three servers in the server group.

Using A records

While the previously described method works under BIND 4, multiple CNAMES for one domain is not a valid DNS server configuration for BIND 9 and above. In such cases, you would use multiple A records as indicated below to effect load balancing via DNS.

```
www.domain.com  IN  A  198.78.46.1

www.domain.com  IN  A  198.78.46.2

www.domain.com  IN  A  198.78.46.3
```

As an alternative to the above, a Time-to-Live (TTL) field is added to A records. The TTL value indicates the maximum time information should be considered to be reliable. By setting a low TTL value, you can ensure the DNS cache is refreshed faster, which will improve load sharing on your organization's Web servers. Unfortunately, the tradeoff of this improvement is the load on your organization's name server will increase. The following example shows using a 60-second TTL value when A records are used.

```
www.domain.com  60  IN  A  198.78.46.1

www.domain.com  60  IN  A  198.78.46.2

www.domain.com  60  IN  A  198.78.46.3
```

Using CNAMES or A records to obtain a DNS round-robin method of load balancing is transparent to clients. Because it uses existing hardware and software, it is very cost effective only requiring a minimum of effort and is used by small and medium sized organizations. Unfortunately, there is no way for DNS to detect the failure of a server, which means when a server fails, DNS will continue to route clients to it. Thus, if there are five servers in your A records and one fails, then 20 percent of client requests will be sent to a server that cannot respond to their requests.

chapter six

The CDN Enterprise Model

The purpose of this chapter is to examine content delivery with respect to the enterprise. Commencing with a discussion of when and why content delivery should be accomplished in-house, we will turn to several techniques that can be employed to facilitate the delivery of content. Because the most effective method to determine an appropriate content delivery mechanism requires knowledge of where data must flow, and the quantity of data in each flow, we need some knowledge of the enterprise's operating environment. This author would be remiss if he did not include a discussion of traffic logs. Now we have a general appreciation concerning where we are headed, let's proceed and investigate the Content Delivery Network (CDN) enterprise model.

6.1 Overview

From a realistic point of view, the size of an enterprise can significantly vary. Some enterprises that host a Web site could be a small retail store. In comparison, other types of enterprises can range in size from Fortune 500 industrial organizations, to large banks and savings and loan organizations. An enterprise model that may be well suited for one organization could be ill suited for a different size organization. Recognizing the fact that it would be very difficult, if not impossible, to discuss a series of CDN enterprise models that could be associated with different types of organizations resulted in this author using a different approach to the topic of this chapter. Instead of attempting to match CDN enterprise models to organizations, a better approach would be to discuss such models in a structured order. The structured order will let readers consider one or more models that could be well suited to the operation of their organization, providing them with the ability to select the most appropriate model. As we examine different CDN enterprise models, we will commence our effort by discussing simple data center models applicable to organizations that operate a single Web site. Then, we will examine the more complex CDN enterprise models that are applicable for organizations capable of operating multiple data centers. For

Table 6.1 Rationale for In-House
Content Delivery

• Concentrated customer base
• Distributed locations available for use
• Knowledgeable staff
• Control
• Economics

the purpose of this chapter, we will use the term data center quite loosely to describe a facility where one or more Web servers are located.

Rationale

Any discussion of using content delivery in-house should commence with an investigation concerning the rationale for doing so. After all, for many organizations that lack the staff, time, or equipment the simple solution is to outsource content delivery.

Concentrated customer base

Table 6.1 lists five key reasons you may want to perform content delivery in-house. The first reason is a concentrated customer base. Having content delivery in-house simplifies the ability to deliver content in a timely manner to that base. A concentrated customer base, such as persons that utilize online ticket ordering for sports events at an arena in Macon, Georgia, or a similar activity that typically originates requests from the immediate geographic area. The operator of the Web site does not have to be concerned with reducing latency from potential Web users originating traffic from areas outside the State of Georgia or outside the United States. The primary focus of such Web site operators should be on maintaining availability to the site. Depending on the traffic volume to the Web site, the site operator may consider operating multiple servers, using a load balancing technique to provide a high level of access to their facility.

Distributed locations available for use

A second reason for considering in-house content delivery is when an organization has distributed locations available for supporting Web servers where it may be possible to create an organizational wide CDN. For example, consider an organization headquartered in Sweden that has branch offices in New York, Singapore, Sydney, Tokyo, London, and Moscow. Suppose that organization currently operates a single Web server in Stockholm and a network traffic analyzer indicates that users accessing the server from Japan and Australia were experiencing significant latency delays. Because the Sweden-based organization has branch offices in Tokyo and Sydney, it may be possible to install servers at those two locations to enhance access from Australia and Japan. With appropriate programming, traffic directed to the Web server located in Sweden would be redirected to a server in either Tokyo or Sydney, based on the origin of the initial data flow to the server in Sweden. Using two distributed locations outside the main office would enable locations that experience significant latency to obtain speedier access to data because Web servers would be located considerably closer to the browser user.

Knowledgeable staff

It makes no sense to develop an in-house content delivery networking capability if an organization lacks personnel to support the network. The availability of a knowledgeable staff or the resources necessary to hire and train applicable personnel is a prerequisite for being able to develop an in-house content delivery networking capability.

Control

A key advantage associated with in-house content delivery versus using a third party is control. By assuming responsibility for content delivery, an organization becomes capable of reacting faster to changing requirements. In comparison, a third party may require a lengthy contract modification to have the service provider initiate one or more content delivery changes to satisfy changing organizational requirements.

Economics

One of the more important reasons for developing an in-house content delivery networking capability is economics. Can an organization save money by performing content delivery operations in-house instead of signing a contract that places the responsibility for content delivery on a third party? The answer to this question can be quite complex because many factors need to be considered beyond a direct dollar and cent accounting. For example, although it may be more expensive for an organization to establish an in-house CDN, the ability to train employees and directly control the operation of the network could represent a rationale for deciding against a pure economic based use of a third party. Although economics are an important decision criteria many times there can be other factors, which when considered, results in a decision that may not make sense from a pure economic perspective. Now we have an appreciation for the main factors that can govern a decision to perform content delivery in-house, we need to ensure that our organization requires establishing a content delivery mechanism beyond operating a single Web server. We need a way to analyze existing traffic flow that will allow us to recognize bottlenecks that could be alleviated by establishing a content delivery networking capability.

6.2   Traffic Analysis

There are two methods an organization can use to determine the need for some type of content delivery networking capability. Those methods are the analysis of Web logs and the use of a protocol analyzer.

Using Web logs

Regardless of the software used to provide a Web server operating capability, each program will have at least one commonality. That commonality is the ability to record Web activity into logs that can be analyzed to provide information about users accessing the server. Because different Web server application programs vary with respect to their capability, we will focus on the type of logs generated by the Apache open source solution for Web site operations.

Apache represents a popular Web site solution that includes a very flexible Web logging capability. In addition to supporting error logging of messages encountered during operations, Apache has the ability to track Web site activity. Apache generates three types of activity logs — access, agent, and referrer. Information recorded into those logs track accesses to your organization's Web site, the types of browsers being used to access the site, and the referring Uniform Resource Locator (URLs) of the site from which visitors arrived. Using various configuration codes makes it possible to capture every piece of information within each HTTP header, providing a significant amount of information about each inbound access request to a Web site. Table 6.2 lists the Apache strings used to log information to the log file, and a short description of the data logged as the result of a particular configuration string.

Using logging strings

To illustrate the potential of the Apache logging strings listed in Table 6.2, let's assume an organization wants to log the remote host, the date and time of the request, the request to the Web site, and the number of bytes transmitted in the request. To accomplish this you would enter the following command:

```
LogFormat "%h %t %r %b" combined
```

The string LogFormat starts the line and informs the Apache program that you are defining a log file type referred to as combined, which generates a single log file instead of individual access, agent, and referrer log files. In addition to being able to create a combined file for a single Web site, Apache supports the ability to create a single log file for large Web sites that have multiple servers. A Web site operator can analyze the data flow to multiple servers located within a physical location or scattered around the globe.

Table 6.2 Apache Logging Strings

String	Description
%a	Remote IP address
%A	Local IP address
%B	Bytes sent, excluding HTTP headers
%b	Bytes sent, excluding HTTP headers, in CLF format
%C	Content of cookies in request sent to server
%D	Time taken to serve the request, in ms
%e	Contents of the environmental variable
%f	Filename
%h	Remote host
%H	The request protocol
%I	Bytes received including request and headers
%i	Contents of header line(s) in the request sent to the server
%l	Remote logname
%m	The request method
%n	The contents of a specified note from another module
%O	Bytes sent, including request and headers
%o	The contents of header line(s) in the module reply
%p	The carnonical port of the server serving the request
%P	The process ID of the child that serviced the request
%q	The query string, if it exists
%r	The first line of the request
%s	The status of the original request
%T	The time required to service the request
%U	The URL path request
%V	The canonical server name of the server servicing the request
%X	The connection status when the response is complete

Web log analysis

Although Apache and other Web application programs can create logs containing predefined activity, such activity represents raw data. To convert the raw data into a meaningful report requires a reporting tool. Although some Web application programs also include a reporting module that when invoked will operate against the log to generate a report, there are times when a Web site manager will require more information. This resulted in the development of a series of application programs and collections of scripts used to generate reports from Web logs. Although this author will leave it to readers to decide which application programs or collection of Web log analysis scripts are most suitable for their operating environment, we can discuss some of the common reports generated by such programs and scripts. This will give us a solid indication of the type of data we can obtain from Web logs and why we need some additional tools to determine if our organization requires a content delivery capability.

Top referring domains

The Top Referring Domains is one of the more important reports provided by application programs and Web log analysis scripts. This report logs the URLs reported by browsers directing them to various Web pages on your server. Typically, application programs and Web log analysis scripts lets the user list the top 10, 25, or another number of the top referring domains.

Through the use of the Top Referring Domains report, you can note which sites are providing the referrals to your organization's Web site. For example, consider an example of the top ten referring domains report included in Table 6.3. From an examination of the entries in Table 6.3 you can note that the primary referring domains represent search engines. Although the two highest referring sites in our example (google.com and

Table 6.3 An Example of a Top-10 Referring Domains Report

Referrals	Domain
127,183	google.com
114,783	yahoo.com
87,413	google.co.uk
67,137	google.ca
47,412	msn.com
46,193	yahoo.de
45,237	google.fr
31,017	yahoo.fr
10,946	asialinks.com
3,105	cnn.com

yahoo.com) are located within the United States, other search engines providing a referring service are located in the United Kingdom (uk), Canada (ca), Germany (de), and France (fr). A majority of referring occurs from Google and Yahoo! search engines located within the United States, but there is also a significant amount of referring occurring from search engines located in Western Europe. Although the referring from search engines located in Western Europe could indicate that the placement of a server outside the continental United States might be warranted, the previously mentioned report does not indicate the delay or latency associated with the referrals. This means that we need additional information prior to being able to reach an intelligent and informed decision concerning the placement of a server outside the continental United States. Unfortunately, we cannot obtain such information from Web logs. However, prior to discussing how we can use a protocol analyzer to obtain the necessary information let's complete our discussion of Web log utilization by examining some of the statistics we can obtain from those logs.

Web log statistics

Because Apache and other Web server products enable you to log most, if not all of the contents of HTTP headers, it's possible to obtain valuable statistics by summarizing the contents of logs. Two of the more important statistics you can obtain from Web logs are access to your organization's server based on origination country and origination time zone.

Origination country

A distribution of access to your organization's Web site by origination country provides you with an indication of the flow of traffic to your server. This information can be helpful for determining the location of server requests and the effect of your organization's advertising and marketing efforts. However, since we are concerned with facilitating content delivery we will leave advertising and marketing concerns to other publications.

Returning to the origination country information, some reports simply list the number of Web page views by origination country in descending order. While this type of report refines raw log data, you will probably have to take pen to paper and use a calculator to further analyze the report. You will more than likely need to group origination countries into areas, such as Western Europe, Eastern Europe, and similar areas to determine if the number of page views that originated from an area justifies the movement of content toward that area. For example, a top countries' list could indicate that between 3 percent and 4 percent of traffic to your organization's Web site originated from countries like Argentina, Chile, and Brazil. Although the traffic originating from each of those countries might not be sufficient to justify establishing a server in an office in South America, cumulatively traffic between 9 percent and 12 percent of total page views could be sufficient.

You may need to periodically take the results furnished by a top countries' report and group page views by country into regional areas.

Originating time zone

A second summary report available from most Web servers groups pageviews according to the time zone of the browser user. The report lists pageviews in descending order based on the time zone of origination. That time zone is in Greenwich Mean Time (GMT) plus or minus an hourly offset. For example, the Eastern time zone in the United States, which encompasses states from Maine to Florida, is GMT – 05:00. In comparison, Moscow is located in GMT + 03:00 while Tokyo is located in GMT + 09:00.

While a Top Time Zone report can provide valuable information, it's possible this report can provide misleading information. A time zone covers an area of the globe between the north and south poles. A large amount of pageviews occurring from GMT – 05:00, which is the Eastern time zone could originate from Florida, New York, or even Argentina, because those locations are all in the same time zone. You need to supplement the report of Top Time Zone pageviews with knowledge about the countries from which the traffic originated.

Other statistics

In addition to the originating country and originating time zone reports, a Web log reporting program or a collection of scripts can be expected to generate a series of statistics. Table 6.4 provides a list of the type of statistics you can typically obtain from a Web log reporting program or commercial script.

In examining the entries in Table 6.4, you will note that the statistical information provides general summary information about the activity occurring at a particular Web site. Although this information may be suitable for examining the need to upgrade an existing Web server, the summary information is not suitable for deciding if an in-house content delivery mechanism

Table 6.4 Web Log Analysis Statistics

Monthly, Weekly, or Daily Statistics
• Total pageviews
• Total visits
• Unique visitors
• Hourly unique visitors
• Average pageviews per hour
• Pageviews per visit
• Busiest day in reporting period
• Busiest hour in reporting period

is warranted based on existing traffic. The originating country report supplemented by an originating time zone report can be used to consider the need for a distributed content delivery capability. In comparison, the summary statistics are better suited for evaluating the capacity and processing power of an existing Web server or group of servers against the current workload request flowing to the server or group of servers. For example, average pageviews per hour provide an indication of the number of pages a server must generate. By dividing that average by 3600, you obtain the number of pageviews per second, which can be compared with the capacity of your server to generate Web pages.

While it's true that not all Web pages have the same composition, and the variance in Web page composition will affect the pages per second (PPS) rate that a hardware platform can generate, working with an average pageview per second and comparing that value to the average server page generation capability provides a reasonable gauge concerning server performance. As the pageview per second value approaches the server's PPS generation capability, the load on the server increases to a point where the server must either be upgraded or an additional server and a load balancing mechanism is required.

Other logging information

Previously we noted that some of the types of data recorded to Web logs are more suitable for investigating server performance than for determining the viability of content delivery. To ensure readers have a solid understanding of server performance, we will conclude our discussion about logs by turning to tools within a server's operating system used to determine the capacity of the hardware with respect to existing and projected workloads. Because Microsoft's Windows operating system represents the dominant operating system used by Web application programs, we will look at the Performance Monitor tool built into Windows to obtain an appreciation for how we can open a window to view the performance of a server's hardware platform.

Microsoft's Performance Monitor

Microsoft developed Performance Monitor to view, log, and chart the values of various performance-related counters. Performance Monitor can spot trends that signify a hardware upgrade or replacement is necessary. Performance Monitor is a tool for examining the ability of the operating system and its hardware platform to satisfy operational requirements. The term Performance Monitor was used by Microsoft to reference this tool when it was bundled with Windows NT. When Microsoft introduced the Windows 2000 server and the Microsoft Management Console (MMC), it changed the name from Performance Monitor to Performance. Because Performance monitors the performance of the computer, this author will refer to this tool using its original name.

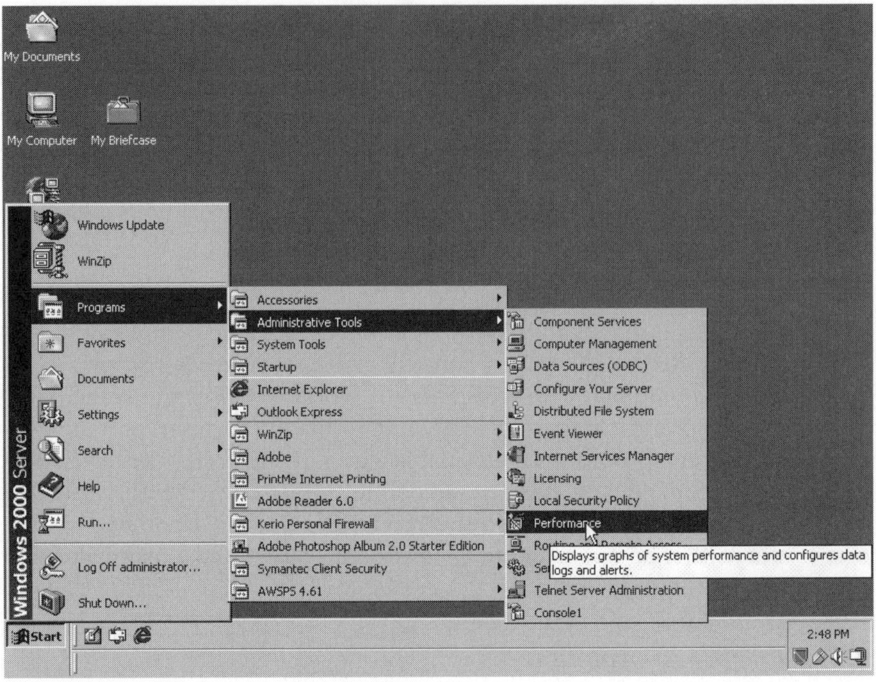

Figure 6.1 Accessing Windows 2000 server's performance monitor provides the ability to display graphs of system performance.

Activating Performance Monitor

Figure 6.1 illustrates using the Start menu under Windows 2000 Server to select the built-in Performance Monitor application bundled with the operating system. As indicated by the sequence of highlighted menu entries, you would select Programs > Administrative Tools> Performance to invoke the Performance Monitor.

Under Windows 2000 server, you access the Performance Monitor via the MMC as illustrated in Figure 6.2. Note that the left portion of the display provides a tree of activities that can be invoked. The right portion of the display provides a graph of the objects you previously selected. After you select one or more objects, the display will show each object in a different color during the selected time period and indicate, for each graphed entry, its latest value (last), average, maximum, and minimum values.

Adding counters and instances

If you examine the lower portion of the graph shown in Figure 6.2, you will note that one counter was previously selected. That counter indicates the percentage of processor utilization, which can be an important indicator concerning the need to either upgrade or replace an existing server platform.

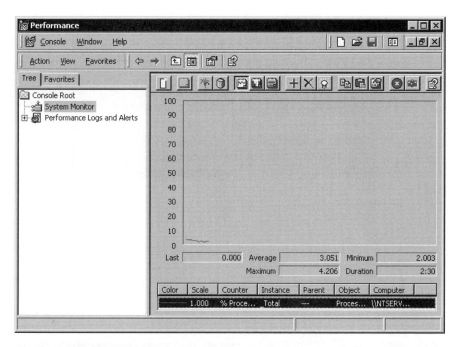

Figure 6.2 Windows 2000 server's performance monitor runs under the Microsoft management console.

In a Microsoft Windows environment, a server's operating system can support hardware with multiple processors. Each processor is referred to as an instance. You can select an applicable counter or group of counters to be plotted by right clicking on the graph, to display a dialog box labeled Add Counters as illustrated in Figure 6.3. Note that Performance Monitor lets you select one or more counters for each instance. In the example shown in Figure 6.3, the % Processor Time counter was selected. Because the computer operating Windows 2000 Server only had one processor, the right area of the display shows Total selected when the button Select Instances from list was clicked. Otherwise, if there were multiple processors the counter could be selected for an individual processor or for all processors.

A second method you can use to select counters to monitor performance is by clicking on the Performance Logs and Alerts entry in the tree portion of the window. Three sub-branches display when you explode the Performance Logs and Alerts branch. Those sub-branches are shown in the left portion of Figure 6.4. In this example, the Counter Logs entry is shown selected, displaying two previously defined logs in the right portion of the window. A dialog box labeled System Overview Properties displays when you right click on either entry or a blank display. That dialog box is shown in the right foreground of Figure 6.4. Note that there are three tabs in the System Overview Properties dialog box. Those tabs are labeled General, Log

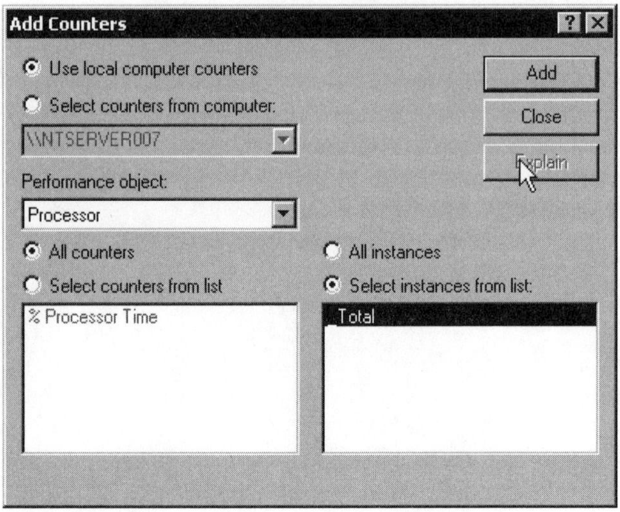

Figure 6.3 System overview properties: the add counters dialog box provides a mechanism to define the counters whose values you wish to monitor.

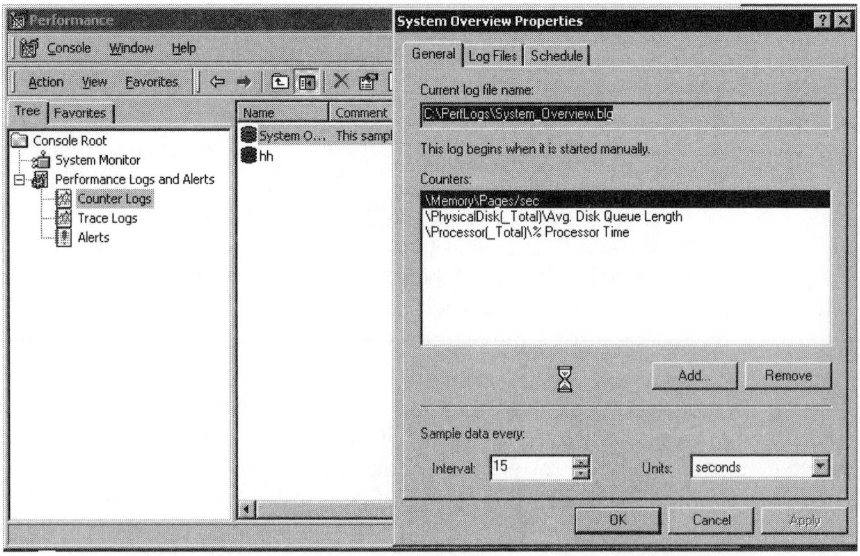

Figure 6.4 Double clicking on a log results in the display of the system overview properties dialog box, showing the counters in the log.

Files, and Schedule with the tab labeled General shown positioned in the foreground.

The General tab indicates the name and location of the current log file. You can add or remove counters and define the sampling interval. The tab labeled Log Files lets you control various aspects of a log file, such as its format, the amount of storage to be used, and the assignment of a comment to the file. In comparison, the Schedule tab lets you define when logging occurs.

Working with Performance Monitor

To better illustrate some of the functionality and capability of the Performance Monitor, let's modify the previously selected log file. You can use the Add or Remove buttons. Because this author wants to add some counters, he will click on the Add button shown in the General tab located in the System Overview Properties dialog box, which displays a Select Counters dialog box displays..

The left portion of Figure 6.5 illustrates the selection of the Add button while the right portion of the display shows the Select Counters dialog box. If you focus on the three buttons in the Select Counters dialog box, you will note a gray colored button labeled Explain, which was selected by this author. Selecting that button displays a textual explanation about each highlighted counter. In the example shown in Figure 6.5, the % User Time counter was highlighted, which displays an explanation concerning what the counter signifies. As indicated by the explanation, % User Time represents the percentage of non-idle processor time spent in user mode. The term user mode references when applications, environment subsystems, and integral subsystems operate but do not require direct access to hardware or all memory. As the % User Time value increases the load on the processor increases.

Once you select the counters to be monitored and select a schedule, you can view the effect of logging via the Performance Monitor's graphing capability. To illustrate the graphing capability, the selected counters shown in

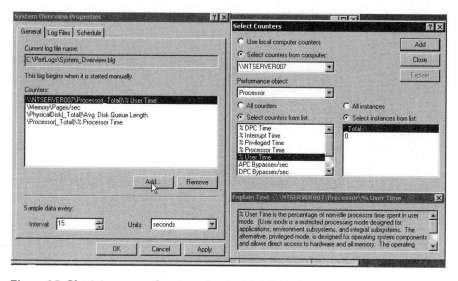

Figure 6.5 Obtaining an explanation about a highlighted counter.

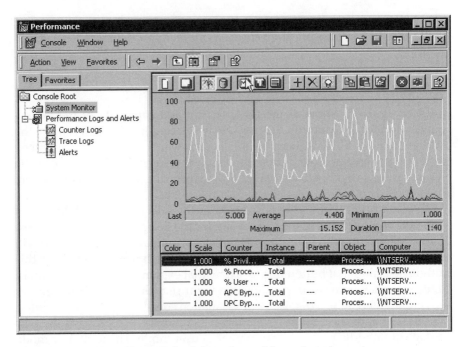

Figure 6.6 Viewing a line graph of the values of five selected counters.

the left portion of Figure 6.5 were plotted as a line chart and as a bar chart. Figure 6.6 illustrates the plot of the selected counters in a line chart format, with the five counters listed at the bottom of the graph. The vertical bar above the 'A' in Average indicates the current time of the plot of the counters, with new plots overwriting the display of previous plots as the vertical line moves from left to right across the display.

To change the composition of a graph, you select an icon above the graph. If you look at the cursor arrow in Figure 6.6, you will note that it is positioned on a line graph icon, resulting in the display of a line graph of the selected counter values. If you move the cursor to the icon to the left, your cursor will be positioned on a bar chart icon. Clicking on that icon will change the graph from a line chart format to a bar chart format as shown in Figure 6.7.

In examining Figure 6.7, note that only the display of the graph changed when a different graph icon was selected. The lower portion of the graph, which contains the color legend for each counter, remained as is.

Summary

Windows Performance Monitor and other tools can be valuable for determining the utilization of hardware, processor, memory, and disk usage.

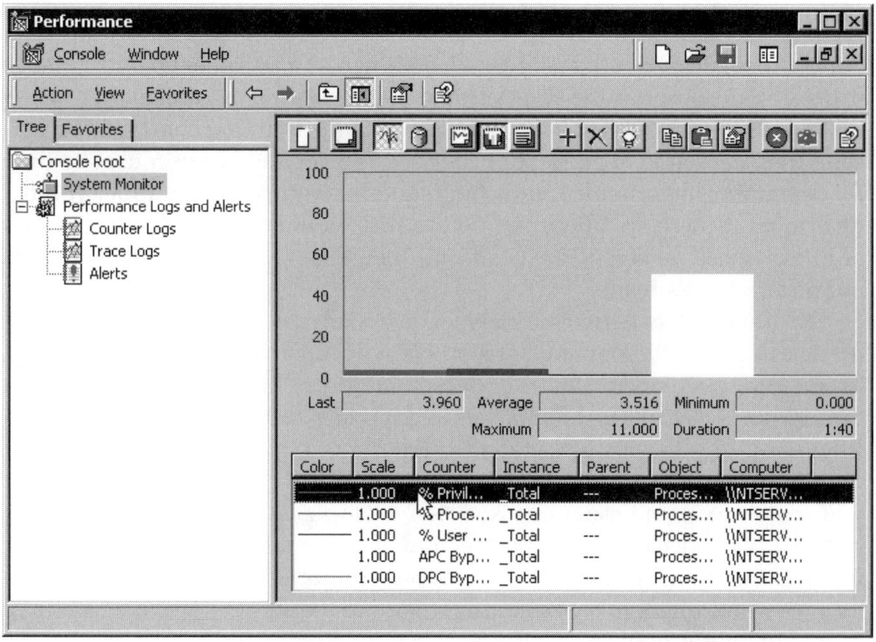

Figure 6.7 Viewing a bar graph of the values associated with five counters.

By carefully monitoring counters associated with the utilization of processors, memory, and disk activity, you can get an insight to potential utilization problems prior to these problems actually occurring. This in turn will provide you with the information to consider several options to improve the content delivery capability of your organization's Web server or servers. These options can include adding processors, memory, and more or faster disk storage to existing hardware, adding servers, or replacing existing hardware with more capable equipment. Regardless of the option you select, having the ability to view key performance metrics provides you with a detailed insight concerning the operation of your organization's server hardware platforms and represents a key tool you can use to facilitate content delivery.

Now that we have an appreciation for the use of server operating system tools, we need to concern ourselves with the flow of data across the Internet. Although Web logs provide a good indication concerning where data is arriving from, these logs do not indicate if browser users are encountering abnormal network delays that would justify the distribution of servers beyond an organization's primary data center. To obtain an insight into the network, we need to turn to a different set of tools. One of those tools is a network analyzer, other tools, such as ping and trace root (tracert) are built into most modern operating systems.

Using a network analyzer

An appropriate network analyzer, also known as a protocol analyzer, makes it possible to observe the flow of data between your organization's server and browser users. That observation can include a timing chart, which indicates the interaction of packet flow between server and browser user by time. By examining the interaction of the requestor and server by time, you can determine if there are abnormal delays that would warrant the installation of one or more servers at areas around the globe to facilitate access to your organization's Web site.

To illustrate a network analyzer, we will use a commercial product marketed under the name Observer. Network Instruments of Minneapolis, Minnesota, developed Observer, which supports both wired and wireless Local Area Network (LAN) analysis at data rates from the 10 Mbps of legacy Ethernet to the 1 Gbps of Gigabit Ethernet.

Similar to other network analyzer products, you use Observer to capture and analyze certain types of packets. For example, assume your organization has a LAN connected to the Internet and operates a Web server connected to its LAN. If you have several workstations connected to the LAN, you could create several filters to record traffic routed to the server instead of all traffic routed to your organization's LAN. You would set up a filter to record traffic to the Internet Protocol (IP) address of the server. If the server supports several applications in addition to operating as a Web server, you could also add a filter to limit the data capture to inbound packets flowing to port 80, which represents HTTP traffic. Filtering an IP address and port number would capture packets flowing to your organization's Web server for Web services.

Once you capture relevant packets, you can perform one or more operations on the captured data. Perhaps the best tool provided by Observer is its Expert Analysis option, which decodes the data flow between two devices, such as a browser user and a Web server.

Figure 6.8 illustrates an example of Observer's Expert Analysis option. In this example, the screen is shown subdivided, with a time scale in milliseconds (ms) used for the subdivision. The left portion of the screen shows the flow of packets from a Web browser to a Web server, while the right portion of the screen shows the response from the server. In this example, the Web browser was assigned the computer name Micron. That computer is shown accessing the MSN Web server whose host address is entertainment. msn.com on port 80.

In examining Figure 6.8, you can observe the three-way initial Transmission Control Protocol (TCP) handshake issued by the Micron computer to the Web server. The initial SYN in the three-way handshake occurs at the top of the time scale, with the server's initial response occurring slightly after 109 ms on the time scale. The server responds with a SYN ACK, which results in the Micron completing the three-way handshake by transmitting

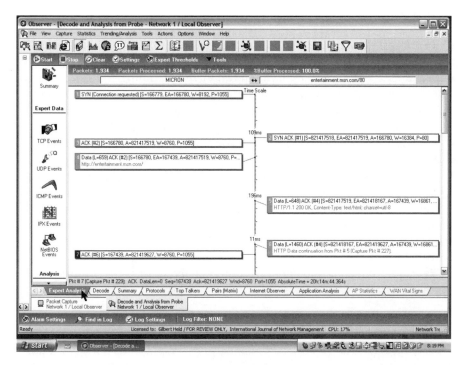

Figure 6.8 Through the use of network instruments observer you can examine the delays associated with the flow of packets between devices.

a SYN packet. After completing the three-way handshake, the Micron computer transmits a data packet further down the time scale, with the server responding at a time slightly after 196 ms on the time scale. For this particular example, the interaction between the client and server are rapid and no substantial network delays would warrant any adjustment to the geographic placement of a server. However, by using Observer on a periodic basis you can examine the data flow between devices and note any abnormal delays that could justify the redistribution of servers to facilitate content delivery.

Other tools to consider

In concluding our discussion of traffic analysis, a word about ping and tracert is warranted to either supplement or replace a network analyzer. If you know the distribution of access originators to your organization's Web site, you can use either tool to determine the delay from your network to a distant device that mimics the location where clusters of remote users are geographically located. For example, assume through the analysis of Web logs, you note that a large number of pageviews to your organization's Web server in Boston occurs from Romania, Bulgaria, and several other Eastern European countries.

While the geographic distance between Eastern Europe and Boston is considerable, if browser users access the Internet via service providers that have a fiber optic link to a major Western European peering point, the transmission delay to Boston may be minimal and not warrant the distribution of content delivery outside the Boston area. Because you would have to either dynamically view Web logs or use a network analyzer in real time to determine the IP addresses of remote users in order to run ping or tracert, a better method is to select a server located within the general area where access requests are initiated to your organization's Web server. Then, you could ping the server to determine the round-trip delay to the distant computer or you could run tracert to examine the delay on each path to the distant server. For either situation, the resulting display will indicate the network associated delay between your organization's Web server and the general area of access from clusters of browser users. Based on the results obtained from the use of ping and tracert and the value of browser user requests to the server, you can then make an informed decision concerning either the distribution of servers to remote sites to enhance content delivery or using a third party to provide a content delivery service.

6.3   Content Delivery Models

In concluding our discussion of content delivery performed by the enterprise, we will examine a series of CDN models. As indicated earlier in this chapter, some models may be better suited to one organization than another may. However, since the number of offices, the activity of an organization's Web site, and the location of browser users accessing the Web site can vary considerably between organizations, our focus will be on enterprise CDN models leaving the selection of an appropriate model to the readers. In this section, we will commence our investigation of CDN models with an elementary single-site, single-server model. Using this model as a base, we will proceed to discuss more complex models. As we discuss each model, we will examine its advantages and disadvantages and cover its operation.

Single-site, single-server model

The simplest method of content delivery over the Internet is the single-site, single-server model. This model results in an organization configuring a single hardware platform at a single location to serve both customers and potential customers.

Advantages

The key advantages associated with the single-site single-server model are cost and simplicity. A single hardware platform reduces both hardware and software costs. Concerning the software costs, because there is only one server, the organization does not have to purchase multiple operating systems or application programs, reducing its software cost to a minimal amount. In addition, the single server requires less support than multiple servers do.

Disadvantages

There are several disadvantages associated with the single-site, single-server content delivery model. The most obvious is the failure of the server, which results in the removal of the presence of an organization from the Internet. Another problem is hardware and software modifications, which could also render the site inoperative. Last, but not least, a single-site requires browser users who may be located around the globe having their requests flow through one or more common telecommunications circuits and peering points to the site. If one or more of those circuits or peering points should become inoperative, a large number of potential users could have their access to the Web site blocked. If a circuit failure occurs within the network of the ISP used by the organization hosting the Web site, it's possible that all access to the site could be blocked even though the server is operational. Thus, the single-site, single-server model results in some key operational deficiencies.

Another problem associated with the single-site, single-server model is all browser access flows to a common server location. This means that browser access from one location could be rapid, with minimal delay. In comparison, browser access from other locations could traverse multiple peering points and have their data flow routed over relatively low speed communications circuits on its way to the central site, resulting in substantial delays. Those delays could result in browser users accessing competitor Web sites or simply abandoning your organization's Web site, resulting in a loss of potential customers.

Single-site, multiple-servers model

A second CDN model retains the single-site location but adds one or more clusters of servers to a single-site, single-server model. Referred to as a single-site, multiple-server model, using multiple servers can usually provide more processing capability than a single server can. In addition, depending on the way multiple servers are connected to the Internet, this model could provide multiple connections to the Internet, increasing the availability of access to the content presented by the organization.

Advantages

Using two or more servers enables the failure of one server to be compensated for by the remaining operational devices. In addition, the effect of hardware and software upgrades can occur on one server at a time, resulting in the other servers in the cluster having the ability to serve customers without having to be placed offline or having their performance degraded.

If multiple connections to the Internet are used, the availability of server content is increased.

Disadvantages

Using a single-site maintains the disadvantages associated with browser users that can be located around the globe accessing a common location. Traversal across many peering points or the routing of data over relatively low speed data circuits can result in significant delays. Such delays can result in browser users abandoning your organization's Web site or accessing a competitor, with either action causing a potential loss of revenue to electronic commerce Web sites.

Although the old adage that two is better than one applies to multiple servers, you need a way to distribute traffic among such servers. This means your organization will require some type of load balancing mechanism to distribute the traffic load in an equitable manner among the servers in your organization's server cluster. Although you can use Domain Name System (DNS) load balancing without any additional cost, as previously noted in Chapter 5, a DNS server has no capability to check the status of other servers in its tables. Thus, one or more servers in a server cluster could fail and, unless modified, the DNS server would continue to refer browser users to the failed server. This means your organization would have to spend additional funds to obtain a load balancing appliance if a DNS server was not an acceptable solution for providing a load balancing capability.

Multiple-sites, single-server per site model

To alleviate potential delays associated with browser users accessing a central site, your organization can consider placing servers at one or more offices. This will result in a new CDN enterprise model involving multiple sites. When a single server is located at each site, we obtain a multiple-site, single-server per site CDN enterprise model.

The primary reason for a multiple-site model will be based on the need to distribute content closer to customers. Assuming you use one or more of the tools previously mentioned in this chapter to determine that a multiple-site model is more appropriate than a single-site model, you need to consider the number of locations where servers should be installed and the number of servers to be installed at each location. The multiple-site, single-server per site model represents the simplest type of geographically distributed content distribution.

Under the multiple-site, single-server per site model, an organization examines actual or potential data flow to its primary Web site. By observing where customers are geographically clustered, and noting delays associated with Internet access from those clustered areas, an organization can determine where one or more servers should be installed outside of the main data center. Thus, the multiple-site, single-server per site represents a way to distribute servers to geographic locations where their installation can enhance browser user access.

Advantages

The primary advantage associated with the multiple-site, single-server per site model is it enables content to be placed closer to browser users. This can reduce latency resulting in a decline in site abandonment, which in turn can result in an increase in customers and customer revenue. A second advantage associated with this model is servers are now placed at two or more distinct locations. Thus, the failure of a communications circuit, an electrical outage, or another type of impairment may not cause an organization's entire Web presence to disappear.

Disadvantages

The comparison of any multiple-site model to a single-site model will result in differences in hardware, software, and support costs. Unless the single-site model employs a large number of clustered servers, the cost of a multiple-site model can exceed the cost of a single-site model.

Another disadvantage associated with the multiple-site model is the way browser users are directed to one of the multiple servers located at geographically dispersed areas. If the direction occurs through the central site, that location becomes a weak link. If that site experiences a power outage or if a circuit linking that site to the Internet fails, then its possible that the redirection facility will fail. You can overcome this problem by setting up separate domains for each distributed server so the distant user requests flow directly to an appropriate server. The key problem associated with this method occurs when content updates are managed from a central site. If that central site should lose its Internet connection, there will be a delay in updating distributed servers. In addition, if the organization maintains its database servers at the central site, which results in the distributed servers having to access the central site, a failure at that location will adversely affect the ability of browser users around the globe to access information. Under this scenario, the solution to the problem is to increase the availability of access to the central site. There are two ways to do this. First, the central site could install redundant communications links to the Internet so the failure of one circuit would be compensated for by a second connection. To provide an even higher rate of network availability, the central site could use two different ISPs. A network problem experienced by one ISP would be compensated for by the ability of data to reach the central site via the second ISP's network.

A second method that can be used to raise the level of availability of the central site is obtained using two or more servers at that location. The CDN model is modified into a new category we refer to as the multiple-site, multiple-server model.

Multiple-site, multiple-server per site model

As its name implies, the multiple-site, multiple-server per site model results in an organization placing two or more servers in at least one location in addition to its central site. The goal behind this model is to provide highly reliable access to content distributed closer to the user. Although this model is both costly and complex to implement, it is used by many multinational organizations that established e-commerce businesses on two or more continents. In addition, there are various versions of the multiple-site, multiple-server per site model that warrant discussion. While the basic multiple-site, multiple-server per site model may imply that initial data flow is directed to a central site, it's possible to set up independent domains. Browser users located in Europe would access a data center located on that continent, while browser users on another continent would be directed to a data center located on that continent. At each data center hosting multiple servers, a load balancing mechanism would distribute the workload over those local multiple servers. In this example, each location functions as an autonomous entity and there is no need for communications between server locations.

The opposite end of the autonomous entity operating model is a non-autonomous entity model. In this situation, each location outside the primary data center communicates with the data center to obtain the dynamic content or other types of information, such as daily price changes. In between the autonomous and non-autonomous entities are partial autonomous entities, where one or more distributed sites only have a requirement to periodically access a central site. Now that we have an appreciation for the potential variances associated with the multiple-site, multiple-server per site model, let's conclude our discussion concerning this model by turning to the advantages and disadvantages associated with this model.

Advantages

There are several advantages associated with a multiple-site, multiple-server per site model. The primary advantage is reliability and availability of access. Using multiple servers per site reduces the probability of a site being down. In addition, because there are multiple locations with multiple servers, it is highly unlikely for the presence of the organization on the Internet to go dark. Because access to each location can be further enhanced by multiple connections to the Internet, it is possible to create a very reliable and highly available presence on the Web while moving content closer to the ultimate requestor.

Disadvantages

Similar to the multiple-site, single-server per site model, the key disadvantages of the multiple-site, multiple-server per site model are cost and complexity. Because multiple servers will be installed at multiple locations, cost can easily become a major issue. In addition, because a load balancing mechanism will be required to distribute the load at each location, the complexity of this model is far above the level of complexity of the other models mentioned in this chapter. When you add development and operational costs, this model is both the most expensive to implement and the most expensive to maintain. However, for some e-commerce multinational organizations the benefits associated with having a direct presence at many locations where Web content can be tailored to the area far outweighs the cost of the effort.

If you travel and access the Internet from several countries, you will probably note you can access the Web sites of major booksellers, home improvement, and appliance stores that are native to the country you are traveling in but, which are branches of multinational firms headquartered in another country. When you access the local Web site, you are accessing either a multiple-site, a single-server per site, or a multiple-site, multiple-server per site Web model.

chapter seven

Web Hosting Options

Until now, we basically subdivided the ability to obtain a presence on the Internet into two categories, do it yourself or employ the services of a content delivery provider. Although the content delivery provider can be viewed as a form of a Web hosting organization, in actuality the term Web hosting can denote a range of organizations to include content delivery providers. Because using a third party for hosting an organization's presence on the Internet represents a content delivery option, we will focus on this topic.

In this chapter, we will discuss the rationale for using a third party Web hosting facility. We will review the major reasons for considering this method of establishing a presence on the Internet. We will discuss the different categories of Web hosting available, the variety of tools provided by third party vendors to facilitate an organization's presence on the Internet, and evaluate factors that separate the suitability of one Web hosting vendor from another with respect to meeting your organization's operational requirements.

7.1 Rationale

There are a variety of reasons an organization will consider when determining if a third party Web hosting arrangement should be used either in place of or as a supplement to an in-house Web hosting arrangement. While cost is normally an important consideration, other factors can have a heavy weight when an organization compares the development of an in-house system to a Web hosting facility. Table 7.1 lists some of the more important factors an organization should consider, with each factor having the ability either by itself or in conjunction with other factors, of forming the rationale for using a Web hosting facility.

Cost elements and total cost

Most Web hosting services bill clients based on several usage-related elements. Those elements can include processor requirements, disk space requirements, total data transfer, and bandwidth utilization. In addition, if

Table 7.1 Rationale for Using a Web
Hosting Facility

- Cost elements and total cost
- Performance elements
- Server side languages supported
- Web Service tools available
- Facility location(s)

your Web hosting service provides an e-commerce site that includes the processing of credit cards, your organization can expect one or more fees in addition to a credit card processing fee.

To ensure a valid comparison between using a third party Web hosting service and an in-house effort, you need to consider the costs associated with these two options. For example, one cost commonly overlooked is electricity, which may be significantly reduced when you use a third party. Instead of paying a direct cost for electricity when your organization operates an in-house Web site, a third party Web hosting facility will normally factor in the cost of electricity, building maintenance, and personnel into their hosting fee structure. Other costs often overlooked are training and maintenance when you operate an in-house Web server. Typically but not always, the savings associated with using a third party Web hosting service will be less than the costs associated with a do-it-yourself approach. However, in spite of economics favoring the use of a Web hosting service, many medium to large organizations will elect the do-it-yourself approach because it provides them with a higher degree of control. This is especially true for the larger e-commerce merchant Web sites, where prices can change very rapidly for certain merchandise and the organization needs to exercise full control over when maintenance and backup operations occur. When the third party Web hosting service operates a shared hosting facility where two or more organizations share the use of a common hardware platform, security considerations need to be examined. Although each organization is provided with a virtual Web server and cannot view or change the operational parameters of the other virtual servers operating on the common hardware platform, organizations are reluctant to use this type of Web service. This is especially true if one or more organizations on the shared hosting platform are competitors. Even though cost can be an important rationale for using a Web hosting facility, by itself it may not be the governing factor.

Performance elements

A performance element is a broad term that references the various factors that govern the operation of the Web hosting facility. Such factors can include the operating system and application software provided by the vendor, the type of hardware platform the Web service will operate on, uptime guarantees for both the hardware platform and communications to and from the

Internet, the Internet connection or series of connections supported by the Web hosting facility, and the type of support the facility provides. In the next few paragraphs we will take a more detailed look at each of these performance elements.

The operating system and applications supported by a Web hosting facility govern your organization's ability to have a particular type of Web site placed into operation. You would not select a vendor that could not provide support of online credit card billing if your organization required that capability. If your organization were a parts' supplier, then the Web hosting facility would need an interface to your organization's mainframe database. Vendors that could not support the required interface would not be considered. If your organization previously developed a Web server that operates under the Sun Unix operating system and you were looking for a hosting site in Australia, you would give preference to hosting sites in that country that can provide Sun UNIX platforms.

The type of hardware platform provided by the Web hosting facility in conjunction with the operating system and application programs will govern the capability of the facility to respond to browser user requests. When examining the hardware platform an organization needs to consider the random access memory (RAM), the number of processors, their operating rate, the data path width, disk controllers, and disk drives. You need to consider the current amount of RAM, if it can be expanded, and, if so, the maximum amount of memory that can be supported. When considering the type of processor used by the hardware platform it's important to note if the platform supports more than one processor and, if so, how many are presently installed. This will provide you with an indication of the upgradeability of the hardware platform. Investigating the capacity of disk drives, current storage used, and the ability to add additional drives will provide you with an indication of the upgradeability of online storage. Because a Web hosting facility typically operates numerous hardware platforms, they may provide more capacity and capability than if your organization acquired a single hardware platform for operation at your facility.

One of the more important performance elements provided by some third party Web hosting organizations is a very high uptime guarantee. Because the Web hosting facility can amortize the cost of backup servers, redundant power, and redundant Internet connections over many users they can usually afford to spend more on maintaining access to servers under adverse conditions than an individual organization can afford. However, not all uptime guarantees refer to the same component. Some uptime guarantees reference Internet availability, with a high uptime guarantee usually associated with sites that have multiple connections to the Internet. Other uptime guarantees refer to the hardware platform. While a Web hosting facility will usually provide higher communications and processor uptime than affordable by small and medium sized organizations, it's important to examine both. Otherwise, a 99.999 percent processor uptime level, which sounds really exciting, may not be all that great if Internet availability is at a much lower level. In this situation, the hardware platform is almost never down, while the Internet connection is problematic.

Two additional performance elements that need to be considered and that usually favor using a Web hosting facility is their Internet bandwidth and their help desk or assistance facility. Because the Web hosting facility provides a service for many customers it more than likely provides more bandwidth and more Internet connections than a single Web site could normally afford. The Internet connections provided by a Web hosting facility, especially if the connections occur via two or more Internet Service Providers (ISP), can provide a higher level of availability than most organizations can afford when constructing an in-house Web site. A Web hosting facility that provides a hosting service for one or more sites with heavy international traffic may operate a 24/7 help desk, a luxury for many small and medium sized organizations.

Server side language support

For most organizations considering using a Web hosting service, the support of server-side languages may be more inclusive then the support from their existing personnel. Although most Web hosting facilities support a core set of server-side languages, they may not support all languages. While a Web hosting facility may provide support for more server-side languages than your organization could support independently, you still need to examine what languages the facility vendor does support. You need to ascertain if all of the languages your Web site will require are supported, and if not, you may have to use an alternative language or barring that, consider a different Web hosting facility whose language support better matches the requirements of your organization.

Web service tools

Another reason for considering using a third party Web hosting facility is the Web service tools they provide to customers. Those tools can considerably vary between hosting vendors. Some vendors may provide Web page construction tools, while other vendors may add one or more promotion tools to their literal bag of tools. Table 7.2 provides a list of major Web hosting facility tools commonly offered and a way to compare those tools against the requirements of your organization. In addition, space is provided for you to compare the offerings of two vendors against your organization's requirements. You can also duplicate the table as a way to compare the tool offerings of additional vendors against the requirements of your organization.

Facility location(s)

The last major factor you can consider to justify using a Web hosting facility is the location or series of locations where the hosting facility resides. Some

Table 7.2 Common Web Hosting Facility Tools

Category/Tool	Organizational Requirement	Vendor A	Vendor B
Site construction tools			
Web templates			
HTML editor			
Java script			
File manager			
Blog builder			
Photo album			
Calendar			
Counters			
URL redirect			
Guestbook			
Email forum			
Message forums			
Password protection			
Chat room			
Audio clips			
Headline news tools			
Today in history			
News headline feed			
Sports headline feed			
Promotional tools			
Daily cartoon			
E-cards			
Links			
Site ring			
Site searches			

Web hosting organizations have several server farms located on different continents while other vendors may be limited to operating a single facility. By carefully examining the location of Web hosting vendors it is possible for your organization to use those facilities to move your content closer to groups of browser users who currently access or have the potential to access your site. For example, assume your organization operates a Web site in Los Angeles and you noted that a large number of page hits occur from browser users located in Western Europe. Further assume that through the use of Web logs supplemented by the use of a network analyzer programmed with appropriate filters, you notice that those browser users are experiencing significant transmission delays, which appear to result in a high level of server abandonment. You might consider using a Web hosting facility located in Western Europe to reduce transmission delays experienced by the browser users accessing your organization's Web site from that general geographic area.

To facilitate traffic from Western Europe flowing to the hosted server, your organization would register a new domain name with a country suffix corresponding to the location of the Web hosting facility. Then, you could modify your existing Web server located in Los Angeles to redirect browser user requests originating from Western Europe to this hosted server. Now that we have an appreciation for the key reasons for examining the potential use of a Web hosting facility, let's turn to the type of hosting facilities available.

7.2   Type of Web Hosting Facilities

There are several ways we can categorize the type of Web hosting facility. Those ways include their ability to support e-commerce, credit card, or Pay Pal usage, the operating system used (Windows, Unix, Linux), geographic location (United States., Canada, Western Europe, etc.), or the manner by which outside customers use Web hosting facilities. The three basic types of Web hosting you can consider include dedicated hosting, shared hosting, and co-location hosting.

Dedicated hosting

A dedicated hosting facility means a server is dedicated to operating a single organization's presence on the Internet. This type of Web hosting should be considered if your organization needs to run customized software and applications instead of standard software provided by the hosting organization. Other reasons where a dedicated server would be preferable include, a need for a high level of security or if your hosted server is expected to receive a high level of traffic, which warrants the use of a hardware platform for the exclusive use of your organization. Because only one organization uses the resources of a dedicated server, the cost will be higher than if two or more organizations share the resources of a single server, referred to as a shared server.

Shared server hosting

A shared server means an organization's Web site operates on a hardware platform that one or more additional Web sites also operate on, in effect sharing the hardware platform. The ability to share the use of a common hardware platform is obtained by a Web server program supporting multiple Web sites. Examples of Web server programs that provide this capability include Microsoft Windows 2000 Server, Windows 2003 Server, and Apache.

This author will turn to Microsoft's Internet Information Services (IIS) to show how multiple servers can be supported on a common Web platform. Under IIS you can right click on the server name (shown with an asterisk in Figure 7.1) established after the program was installed. As illustrated in Figure 7.1, this action displays a pop-up menu, which includes the option New. Selecting the New option lets you define either a new Web site or a new FTP site.

A Web site creation wizard displays to help walk you through the creation process associated with defining a new Web site. The wizard will

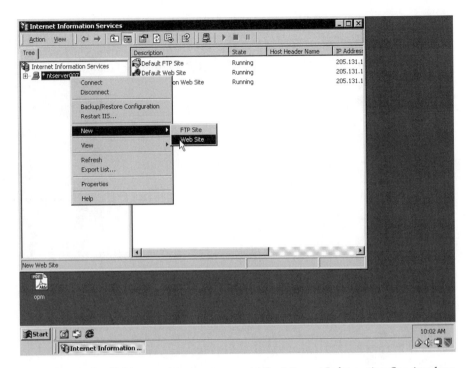

Figure 7.1 Right clicking on the server name in the Internet Information Services box enables you to create a new server.

request you to enter a description of the site, its Internet Protocol (IP) address, and the port field to be used. Next, you can enter the domain name for the site. Assuming we simply entered the name "site 1" for the new Web site, it would appear in the ISS box as shown in Figure 7.2. Thereafter, we could right click on a Web site entry to add another site onto the common hardware platform.

Co-located hosting

A third type of Web hosting arrangement is referred to as co-located hosting. Under a co-located hosting solution, your organization purchases a Web server, which is located at the third party facility. The third party is responsible for physically housing the server, providing power, environmental controls, security, support, and Internet connectivity.

The goal of a co-located hosting arrangement is it provides an organization with the ability to obtain required hardware that may not be available from the hosting organization. In addition, it also provides the ability of an organization that has one or more excess hardware platforms, to have them used by the hosting facility instead of having to be billed for additional hardware usage. Unfortunately, this arrangement can create problems if the hosting organization is not familiar with the hardware. In addition, selecting

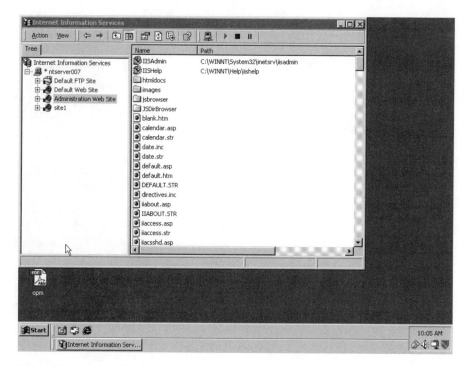

Figure 7.2 Viewing IIS after another Web site was added to the common hardware platform.

a customized hardware solution can create backup problems. Using a co-located hosting method needs to be carefully analyzed based on the hardware you select and the hosting organization's ability to support it.

7.3   Evaluation Factors

In concluding our discussion of Web hosting options, it's important to consider the manner by which you evaluate this option against the offerings of different vendors. If you type in the term "Web hosting" in Google, Yahoo!, or another search engine you will be inundated with responses to view. Google alone provides over 26 million responses.

To facilitate the consideration of a Web hosting organization, we need to examine a large number of quantifiable metrics and obtain answers to questions that may be rankable but may not be quantifiable. A number of quantifiable metrics were previously noted in this chapter. For example, Table 7.2 includes a list of Web hosting facility tools you can consider by matching your organization's requirements against the comprehensive list of tools presented in the table. Table 7.3 contains a list of Web hosting evaluation factors that includes a reference to Table 7.2 for comparing hosting facility tools. Table 7.3 provides a list of features you may wish to evaluate, followed by three blank columns below the column heading. You can specify

Table 7.3 Web Hosting Evaluation Features

Category/Feature	Organizational Requirement	Vendor A	Vendor B
Internet connection			
Bandwidth	_____	_____	_____
Type of connection	_____	_____	_____
T1	_____	_____	_____
T3	_____	_____	_____
Other	_____	_____	_____
Redundancy	_____	_____	_____
Up time guarantee	_____	_____	_____
Type of server			
Dedicated	_____	_____	_____
Shared	_____	_____	_____
Co-hosted	_____	_____	_____
Server operating system			
UNIX	_____	_____	_____
LINUX	_____	_____	_____
Sun OS	_____	_____	_____
Windows 2000	_____	_____	_____
Windows 2003	_____	_____	_____
Other	_____	_____	_____
Server capacity			
Processor speed	_____	_____	_____
Number of processors	_____	_____	_____
Online storage	_____	_____	_____
Tape storage	_____	_____	_____
Other	_____	_____	_____
Server software			
Apache	_____	_____	_____
Microsoft IIS	_____	_____	_____
O'Riley	_____	_____	_____
Other	_____	_____	_____
Server side software			
ASP	_____	_____	_____
C++	_____	_____	_____
Jscript	_____	_____	_____
Perl	_____	_____	_____
PHP	_____	_____	_____
Microsoft SQL	_____	_____	_____
VBScript	_____	_____	_____
Other	_____	_____	_____
Server security			
Certificates	_____	_____	_____
Authentication	_____	_____	_____
type	_____	_____	_____
Firewall	_____	_____	_____
Router access lists	_____	_____	_____
SSL	_____	_____	_____
Other	_____	_____	_____

Table 7.3 (continued) Web Hosting Evaluation Features

Category/Feature	Organizational Requirement	Vendor A	Vendor B
Server backup			
Content backup	_____	_____	_____
On change	_____	_____	_____
Daily	_____	_____	_____
Weekly	_____	_____	_____
Other	_____	_____	_____
Redundant power	_____	_____	_____
Up-time guarantee	_____	_____	_____
Other	_____	_____	_____
Web hosting facility tools			
(See Table 7.2)			
Server statistics			
Events logged	_____	_____	_____
Other	_____	_____	_____
Service and support			
Types of customers	_____	_____	_____
Opinions of customers	_____	_____	_____
Stability	_____	_____	_____
24X7 technical support	_____	_____	_____
Help desk	_____	_____	_____
Other	_____	_____	_____

your organization's requirements for a specific feature in the second column. The third and fourth columns labeled Vendor A and Vendor B, let you compare the offerings provided by two vendors to the features required by your organization. You can make additional copies of the table that will allow you to easily compare the offerings of additional vendors against your organization's requirements.

Although most of the entries in Table 7.3 are self-explanatory, a few words of discussion are warranted with respect to service and support. For many organizations it's important to ascertain the types of existing customer sites being hosted by a Web hosting facility. Obtaining this information will provide a valuable insight concerning the ability of the third party to host your organization's Web site. For example, if the hosting facility you are considering currently operates Web sites for a number of clients that require credit card authorization and credit card payment operations, and your organization anticipates performing e-commerce using credit cards, then the hosting facility should be familiar with the process required. However, to ensure they are, you should ask for references and obtain the opinions of existing customers concerning the performance of functions your organization will require now and in the future.

Index

A

access delays, 74–77
Active X, 36
Address Resolution Protocol (see ARP)
Akamai network, 96–105
Apache, 134–138
application caching, 111–112
application servers, 40–42
ARP, 61
ASP.NET, 111–112
authentication, 124

B

backup, 157
bandwidth, 158
Belgian National Internet Exchange (see BNIX)
BNIX, 90–91
browser cache, 108
browser programs, 32

C

cache-control extensions, 119–121
cache-control header, 116–120
caching, 10,107–121, 124–125
CDN model, 69–105
CGI, 39
client-to-mainframe data flow, 3–5
client-server computing, 3, 5–7
client-server models, 21–46
Common Gateway Interface (see CGI)
Company perception, 72
Content Delivery Network
    advantages, 1
    client-server operations, 16–17

definition, 1
disadvantages, 2
evolution, 2–3
peering point, 17–20
enterprise model, 131–153
models, 148–152
Cookies, 30–31, 125
Co-located hosting, 161–162
Crawling, 15
Customer base, 132
Customer loyalty, 71

D

data link layer, 48
dedicated hosting, 160
distributed locations, 132–133
DNS, 58–67
DNS load balancing, 126–130
DNS resource records, 64–67
Domain name servers, 58–59
Domain Name System (see DNS)
Dynamic ports, 57

E

economics, 133
edge operations, 95
edge servers, 82, 97–100
Edge Side Includes (see ESI)
Egress delays, 78–81
Electricity, 156
Entry considerations, 73
ESI, 100–105
ESI for Java, 103–104
Euro-Ix, 86
Expires header, 116